Anaerobes and Methanogens

Editor

Bhagwan Narayan Rekadwad

Division of Microbiology and Biotechnology
Yenepoya Research Centre
Yenepoya (Deemed to be University)
Mangalore, Karnataka, India

CRC Press
Taylor & Francis Group
Boca Raton London New York

CRC Press is an imprint of the
Taylor & Francis Group, an **informa** business
A SCIENCE PUBLISHERS BOOK

Cover credit: Image obtained by Dr. Bhagwan Narayan Rekadwad from CCC RightsLink under License Number 5994140583449.

First edition published 2026
by CRC Press
2385 NW Executive Center Drive, Suite 320, Boca Raton FL 33431

and by CRC Press
4 Park Square, Milton Park, Abingdon, Oxon, OX14 4RN

CRC Press is an imprint of Taylor & Francis Group, LLC

Library of Congress Cataloging-in-Publication Data (applied for)

ISBN: 978-1-032-79854-7 (hbk)
ISBN: 978-1-032-82904-3 (pbk)
ISBN: 978-1-003-50689-8 (ebk)

DOI: 10.1201/9781003506898

Typeset in Times New Roman
by Prime Publishing Services

Preface

The book will expose the fascinating world of anaerobic microorganisms, organisms that thrive in the absence of free, aerobic form of oxygen. These microbes have evolved diverse strategies to survive and prosper in conditions lacking, or having limiting amounts of oxygen. From methane-producing methanogens to anaerobic bacteria, these organisms are important ecologically in a variety of environments-from most extreme to human-engineered systems.

These chapters will sweep across a wide variety of topics, from the physiology and molecular adaptations of the anaerobes to their taxonomic classification and evolutionary history, including the importance of methane production in global carbon cycling to the potential of harnessing the anaerobic microbes for bioenergy production. This book will also have a special focus on the use of metagenomics in understanding anaerobic microbial communities and the exciting area of micro-aeration, where small quantities of oxygen are added to anaerobic systems. In this way, I hope this book has shown the important role anaerobic microorganisms that exists around us in our surrounding environment.

Bhagwan Narayan Rekadwad

Contents

Chapter 1

Anaerobes and Methanogens

Microbial Pioneers of Oxygen-Free Environments

Abhishek Chauhan,[1,*] *Jayati Arora*,[2] *Ritu Chauhan*,[3] *Pallavi Bhardwaj*,[2] *Anuj Ranjan*,[1] *Rupesh Kumar Basniwal*,[4] *Rajpal Srivastav*[5] and *Tanu Jindal*[1]

1.1 Introduction

The absence of oxygen in the environment is the main requirement of anaerobic microorganisms. Unlike aerobic microorganisms, anaerobes do not require oxygen for their growth, and, in many cases, oxygen can be toxic to them. Various biochemical pathways enable these members of the anaerobic community to generate energy through routes that do not rely on oxygen as the terminal electron acceptor. Instead, these viable alternatives can be employed through anaerobic metabolism to transfer electrons to substance acceptors, such as nitrate, sulphate, carbon dioxide, and organic products, which, in turn, act as a terminal for the metabolic process (Madigan et al. 1997).

Anaerobic environments can be found in diverse habitats, from the interior of the Earth's crust to the intestinal systems of animals. One of the most famous groups of anaerobic microorganisms that have been discovered is the methanogens which

[1] Amity Institute of Environmental Toxicology, Safety and Management, Amity University, Noida 201313, India.
[2] Amity Institute of Environmental Science, Amity University, Noida 201313, India.
[3] Department of Biotechnology, Graphic Era Deemed to be University, Dehradun 248002, Uttarakhand, India.
[4] Amity Institute of Advanced Research and Studies (M&D), Amity University, Noida 201313, India.
[5] Amity Institute of Biotechnology, Amity University, Noida 201313, India.
* Corresponding author: akchauhan@amity.edu

DOI: 10.1201/9781003506898-1

are part of the domain Archaea. An action of methanogens on complex organic materials in an oxygen-free environment generates methane as a byproduct. On the other hand, the methanogenesis process adds greenhouse gas in the environment which may accelerate the climate change problem (Thauer et al. 2008).

Anaerobes are generally categorized into two major categories: obligate anaerobes and facultative anaerobes. Obligate anaerobes are organisms that cannot live and reproduce in the presence of oxygen. Exposure to oxygen can cause oxidative damage to their cellular components, leading to cell death. To detoxify reactive oxygen species in the smallest amounts they may come into contact with these bacteria that have evolved specialized enzyme systems, such as catalase and superoxide dismutase (Van de Werken 2008).

Moreover, facultative anaerobes are more adaptable than their counterparts. In other words, they can grow in both the presence and absence of oxygen. In aerobic conditions, facultative anaerobes utilize oxygen for respiration (a process that is highly efficient and helps them to make ATP). Nevertheless, when there is a lack of oxygen, they alter their metabolism by stopping the anaerobic metabolic pathways such as fermentation or anaerobic respiration, the paths allowing them to continue generating energy and maintaining (Todar 2006).

Besides their ecological functions, anaerobes are of great interest in medical, industrial, and environmental domains. Anaerobic bacteria are usually present in the human body under normal conditions and only occasionally become pathogenic, initiating the development of infections that are highly resistant to oxygen treatment and therefore require specific treatments (Brook 2007). In the manufacturing sector, anaerobic chemical processes are utilized in the treatment of wastewater, the generation of biogas, and the fermentation of different kinds of products such as alcohol, and oxide components (Gerardi 2003).

To acquire deep insights into microbial communities and ecosystem function, it is unavoidable to learn about the biology and ecologies of anaerobes, especially those that thrive in oxygen-limited environments. As the first colonizers of oxygen-free environments, anaerobes, among others, such as methanogens, are the major actors in the biogeochemical cycles and that is why they are important for the creation of stable and sustainable places for them.

1.1.1 Importance of Anaerobic Environments

In many natural and artificial systems, anaerobic (oxygen-free) conditions are essential. These environments are characterized by unique microbial communities that carry out important ecological tasks, support biogeochemical cycles, and provide major advantages for industrial uses.

Anaerobic environments are the main ecological niche in which anaerobic microorganisms are involved in such nitrogen, carbon, and sulphur cycles. The microorganisms that do not require oxygen to survive including methanogens, sulphate bacteria, and denitrifiers contribute to the release, decomposition, and transformation of organic and inorganic materials by microbial activity undergo them. Methanogens are live things that perform the process by which methane is produced (methanogenesis). Methanization is a biological method that not only recycles carbon in ecosystems but is also concerned with the regulation of atmospheric methane, the potent greenhouse gas (Conrad 2007).

In the nitrogen cycle, anaerobic environments are also able to enable processes such as denitrification and the process of nitrogen and ammonia being converted to nitrogen gas through anammox (anaerobic ammonium oxidation), therefore, nitrogen pollution in the surrounding area is reduced (Jetten et al. 2001). Via the sulphur cycle as well, sulphate-reducing bacteria are the ones to make sulphates valuable resources by converting them to hydrogen sulphide, which in turn, causes these substances to be made available to and transported in sediments and soils (Muyzer and Stams 2008).

Anaerobic conditions are crucial as well as in the health and function of the natural ecosystem. The wetlands are one of the most astonishing anaerobic environment ecosystems that give many important functions such as water purification, regulation of floods, and carbon sequestration. The lack of oxygen in the wetlands allows these microbial processes to be very helpful for absorbing pollutants and for this reason, the carbon remains in the peat even when the climate is changing (Mitsch and Gosselink 2015).

Concerning human health, anaerobic environments are important, since they are inhabited by good anaerobic bacteria in the human gut. Typically, these are bacteria that break down the food in the stomach, assimilate nutrients, and protect the digestive tract. They breakdown the complex food into simple pieces, and then, they use the glucose to make other necessary substances like amino acids, lipids, etc., and maintain gut health by protecting against pathogenic microorganisms (Louis et al. 2014).

Industrial applications of anaerobic environments are diverse and impactful. Among the largest, anaerobic digestion is the most prominent. This is the process used to dispose of organic waste and make biogas, a renewable energy source mainly made up of methane. This technology is often being employed in wastewater treatment plants and agricultural operations to dispose of waste, reduce greenhouse gas emissions, and generate energy (Appels et al. 2008).

Moreover, anaerobic fermentation plays a vital role in both biotechnology and the food industry since it is the very process through which these industries can produce alcohol, organic acids, and antibiotics. According to Kent (2013), it is the process of using the metabolic capabilities of anaerobic microorganisms which metabolize substrates in the absence of oxygen to produce valuable products.

Consequently, anaerobic environments are one of the key elements in nature and human-built systems. They are the key to which the simple and green biogeochemical cycles run, they also support biodiversity by being able to live in environments where oxygen is not a major constituent of the habitat, they also help in climate regulation besides they are the ones that get involved in waste management and renewable energy production.

1.1.2 Introduction to Methanogens

Methanogens are a unique group of microorganisms placed within the domain Archaea that can produce methane as a metabolic end product. They are important in the global carbon cycle and widely distributed in anoxic environments containing very low amounts of oxygen. The production of methane by methanogens of the archaean domain is called methanogenesis. This is important not only for the recycling of organic matter in ecosystems but also for methane being a potent greenhouse gas for climate change. Methanogens are commonly found in wetlands, rice fields, cow rumen, and marine sediments, and also in human intestines. Anaerobic conditions are well adapted by methanogens and it is well supported by using carbon dioxide, hydrogen, acetate, and other one-carbon compounds as substrates for methane synthesis to the bacteria. This biochemical adaptability is enabled by special enzymatic pathways that are characteristic of methanogens, whereby they convert these substrates into either methane and water or methane and carbon dioxide (Garcia et al. 2000). There are several genera of methanogens, each of them special in different environmental niches. Among the best-described genera are Methanobacterium, Methanococcus, Methanobrevibacter, and Methanosarcina. These genera show a variety of morphologies from rods to cocci and live in habitats with strongly differing temperature regimes—from environments that are thermophilic, such as hydrothermal vents, to mesophilic settings like the gut of ruminants. Most methanogens degrade organic matter in anaerobic environments, they play a very critical role. In environments such as wetlands and peat bogs, methanogens flourish in the final stage of decomposition and convert multifarious organic compounds to methane. Apart from nutrient cycling, the process also helps in determining the carbon balance within these ecosystems. From a global warming perspective, methane that comes from wetland environments is very significant because it has been recognized that the global warming potential of methane per molecule is 25 times more than carbon dioxide over 100 yr (Change 2007). The great ecological importance of methanogens is found in biotechnological fields besides their recognition. They are indispensable in the anaerobic digestion process and are utilized for wastewater treatment and biogas production (ElMekawy et al. 2013). Methanogens contribute to the conversion of organic wastes into biogas, a renewable source of energy, whose active component is methane. This biotechnological application hence not only involves sustainability in waste management but also energy generation, therefore, replacing dependency on fossil-based fuel (Schmidt et al. 2016). Moreover, methanogens have been examined for their application in bioremediation and biofuel production. The fact that methanogens can carry out substrate-level fermentation under anaerobic conditions has presented them as potential microbes to be used in treating industrial effluents and harnessing energy from waste materials. In summary, methanogens are a vital group of microorganisms that make key contributions to both natural and engineered anaerobic environments. Their unique metabolic potential drives several key ecological processes and opens promising applications in the field of sustainable energy production and waste management. The understanding and potential harnessing of methanogen potentials are paramount not only for advancement in microbial ecology but also for the innovation of new biotechnological solutions.

1.2 Evolutionary History

1.2.1 Origins of Anaerobic Life

Anaerobic life probably came to exist from the earliest days on Earth, way before the presence of oxygen in the atmosphere. Formed about 4.6 billion *yr* ago, the Earth initially had an atmosphere with very little oxygen but was rich in nitrogen carbon dioxide, methane, and ammonia. Life on Earth began in this primordial environment about 3.8 billion *yr* ago, these first organisms were strictly anaerobic microorganisms, indicating that they lived in an atmosphere depleted of oxygen. Very early life on Earth was largely anaerobic in metabolism: methanogens, sulphate-reducing bacteria, and fermenters constituted the majority of the microorganisms that developed metabolic pathways that served to derive energy from both organic and inorganic molecules, all in the absence of oxygen. Methanogens, such as taking in hydrogen and carbon dioxide to give methane, a metabolic reaction that has never stopped since then and is part of some ecologies today (Ueno et al. 2006). Stromatolites are, in fact, the earliest available fossil records that demonstrate evidence for anaerobic life and are comprised of those laminae that are produced by microbial communities. Such structures, dated at 3.5 billion years back, would imply the domination of ancient shallow marine environments by microbial mats, characterized by anaerobic bacteria. The metabolic activities of the microbes were henceforth an important factor in shaping the chemical composition of early oceans and the atmosphere (Allwood et al. 2006). Probably, a key step in evolution was the development of photosynthesis, and more particularly anoxygenic photosynthesis, since in this case, no oxygen is a byproduct. That process was driven likely by early phototrophic bacteria, among them green sulphur bacteria and purple sulphur bacteria, which used light energy to fuel the synthesis of organic compounds. These bacteria lived in the photic zones of ancient aquatic environments where light was present, but oxygen was not (Blankenship 2010). Oxygenic photosynthesis first appeared approximately 2.4 billion yr ago with cyanobacteria, it marks the beginning of the Great Oxidation Event. This major event finally led to the accumulation of oxygen in the atmosphere and oceans, which made irreversible changes in the redox state of Earth and, therefore, made it possible for the evolution of aerobic life forms. However, simultaneously with the establishment of anoxic conditions on Earth, anaerobic organisms were able to colonize anoxic niches for their life—in particular, deep sediments, swamps, and animal guts, where they survive and to the present day play a crucial role in biogeochemical cycles (Lyons et al. 2014). The persistence of anaerobic life throughout Earth's history highlights its adaptability and evolutionary importance. Sulphate-reducers, fermentative bacteria, and modern methanogens, as a subset of the main groups of modern anaerobes, have a common ancestry with these ancient lineages and continue to play critical roles in today's ecosystems. Their evolutionary history yields insight into the resilience and diversity of life under changing environmental conditions. In conclusion, the roots of anaerobic life are firmly embedded in Earth's early anoxic setting. Anaerobic microorganisms are not only the most ancient forms of life, but they have persisted throughout billions of years of evolution. The metabolic novelties of these microbes set the stage for the complex interplay of life and geochemistry that characterizes our planet.

1.2.2 Adaptations to Oxygen-Free Environments

Anaerobic microorganisms have developed a variety of different adaptations to allow them to live in the absence of oxygen. These are mainly related to their metabolic pathways, cellular systems, and ecological approaches that cope with the lack of oxygen.

1.2.2.1 Metabolic Pathways

The most important adaptation is the generation of other metabolic pathways, the last steps of which do not use oxygen as an electron acceptor. Anaerobes can oxidize nitrate, sulphate, carbon dioxide as well as organic compounds. For instance, methanogens reduce CO_2 through the use of hydrogen gas to produce methane in a process known as methanogenesis (Thauer et al. 2008). Sulphate-reducing bacteria reduce sulphur and then release hydrogen sulphide from sulphate (Muyzer and Stams 2008).

1.2.2.2 Fermentation

Anaerobes primarily utilize fermentation to produce energy. In fermentation, the organic substrates are broken down into simpler compounds like acids, alcohols, and gases without the involvement of oxygen. Through this process, anaerobes can produce ATP through substrate-level phosphorylation. Various ways can lead to the conduction of fermentation, including lactic acid, alcoholic, and mixed-acid fermentations, all these may vary according to different niches in environments (Madigan et al. 1997).

1.2.2.3 Enzymatic Detoxification

Although anaerobes thrive in environments devoid of oxygen, free oxygen is usually prevalent in low amounts. As such, most anaerobes have developed a suite of detoxification enzymes against Reactive Oxygen Species (ROS), which include superoxide dismutase, catalase, and peroxidase. These enzymes protect cells from oxidative stress by allowing the cells to survive in the face of brief exposure to oxygen (Blokhina et al. 2003).

1.2.2.4 Cellular Structures

Anaerobes also have specialized cellular structures that make them better adapted to anoxia. A good example is some methanogens, whose cell membrane lipids have unique structures that give them high stability and functionality in extreme environments. The presence of ether-linked lipids instead of the ester-linked lipids found in bacteria and eukaryotes provides more resistance to both hydrolytic and oxidative damage as previously alluded to (Koga and Morii 2007).

1.2.2.5 Ecological Strategies

Anaerobes often adopt symbiotic relations with other micro-organisms as a means of enhancing their life opportunities. In such consortia, metabolic end products of one species represent substrates of another, and tight webs of metabolic interdependencies arise. For example, fermentative bacteria in syntrophic association with methanogens degrade complex organic compounds in wetlands and the rumens of ruminants. These habitats are very effective natural bioreactors (Schink 1997).

1.2.3 Role in Early Earth's Atmosphere

Anaerobic microorganisms were significantly shaping early Earth's atmosphere, more so during the Archean eon, which was between about 4 to 2.5 billion year ago. During this period, the atmosphere was mainly anoxic and was composed mainly of nitrogen, carbon dioxide, methane, and other gases, with little free oxygen. The most important role played in driving the biogeochemical cycles that, in turn, influenced the composition of the early atmosphere was that of anaerobic microbes, and this included methanogens, sulphate-reducing bacteria, and other anaerobic fermenters.

1.2.3.1 Methanogens and Methane Production

Among the major methane producers during the Archean were methanogens, which form a group of anaerobic archaea. By using hydrogen and carbon dioxide for methanogenesis, methanogens contributed to the methane content of the primitive atmosphere. Had methane—a potent greenhouse gas—been present, Earth's surface temperature could still have been maintained at habitable levels despite the faintness of the young Sun. It would have ensured that there was a proper greenhouse effect to maintain liquid water and microbial life on its surface.

1.2.3.2 Sulphate-Reducing Bacteria and Sulphur Cycle

These sulphate-reducing bacteria played a major role in the early oceans on Earth during the sulphur cycle. These microorganisms used sulphate as an electron acceptor and reduced it to hydrogen sulphide. This not only impacted the sulphur chemistry of ancient oceans but also influenced the atmospheric composition. The produced hydrogen sulphide and additional reactions with iron and other metals produced metal sulphide deposits, sulphur sequestration, and perturbations in the geochemical cycles (Canfield 2004).

1.2.3.3 Anaerobic Fermenters

A decomposition of organic matter of anaerobic fermenters in the absence of oxygen gave off several gases, methane, hydrogen, and carbon dioxide. Such decomposition is critical to nutrient recycling and sustenance of the carbon flux in the biosphere. The metabolic byproducts of such anaerobic processes could have contributed to the evolving composition of the atmosphere setting the stage for further biological and chemical transformations that were to follow (Kasting 2005).

1.2.3.4 Transition to Oxygenated Atmosphere

The anaerobic activities of microorganisms in the past marked the onset of oxygenic photosynthesis by cyanobacteria that occurred about 2.4 billion years ago and were the source of the Great Oxidation Event (GOE), starting to accumulate substantial levels of oxygen. While the produced oxygen initially was rapidly scavenged through reactions with reduced materials like iron and methane, it held off the accumulation of free oxygen within the atmosphere. However, anaerobic life forms set the stage for a later increase in atmospheric oxygen levels, a shift in planetary redox state that set the stage for the expansion of aerobic life (Lyons et al. 2014).

1.3 Anaerobic Metabolism

1.3.1 Overview of Anaerobic Respiration

Anaerobic respiration is a metabolic process through which microorganisms obtain energy in an environment that lacks free oxygen. Anaerobic respiration depends on the reduction of substrates through electron acceptors other than molecular oxygen. These include nitrate, sulphate, carbon dioxide, and organic compounds (Gerardi 2003). This metabolic ability allows anaerobes to thrive in a diversity of anoxic environments, from deep-sea sediments to wetlands and animal intestines.

1.3.1.1 Electron Acceptors and Energy Yield

Anaerobic respiration makes use of diverse electron acceptors that are dependent on microbial species and environmental conditions, such as:

- **Nitrate (NO_3^-):** Reduced to nitrogen gas (N_2) or ammonium (NH_4^+) by denitrifying bacteria.
- **Sulphate (SO_4^{2-}):** Reduced to hydrogen sulphide (H_2S) by sulphate-reducing bacteria.
- **Carbon dioxide (CO_2):** Reduced to methane (CH_4) by methanogens.

Anaerobic respiration has a lower energy yield compared to aerobic respiration because the alternative electron acceptors are of lower reduction potential than oxygen. However, anaerobic respiration is an effective strategy in oxygen-depleted environments, and can thus sustain the growth of numerous microorganisms.

1.3.1.2 Metabolic Pathways

Anaerobic respiration involves several critical processes:

- **Glycolysis:** This process represents the first phase of glucose breakdown via pyruvate formation resulting in ATP and reducing equivalents (NADH).
- **Fermentation:** Organisms regenerate NAD^+ by reducing pyruvate to fermentation products such as lactate, ethanol, or acetate in the absence of any external electron acceptor.
- **Electron Transport Chain (ETC):** Electrons from NADH and $FADH_2$ are transported through membrane-bound carriers to the terminal electron acceptor, the transfer generates a proton motive force that will drive ATP synthesis by ATP synthase.

1.3.1.3 Ecological and Industrial Importance

Anaerobic respiration plays a major role in nutrient cycling in ecosystems, mainly anoxic ones. It takes place during denitrification and sulphate reduction processes (Lecomte et al. 2018). Anaerobic respiration is also very important in industries where the anaerobic digestion for waste treatment is used to produce biogas in an environmentally friendly manner.

1.3.2 Fermentation Processes

Fermentation is a biochemical process where microorganisms generate energy from organic substrates in the absence of oxygen. It breaks down these substrates into simpler compounds through substrate-level phosphorylation, producing ATP without using an electron transport chain or external electron acceptors.

1.3.2.1 Types of Fermentation

1. Lactic Acid Fermentation:
    - Performed by lactic acid bacteria and muscle cells, this process converts glucose into lactic acid, regenerating NAD^+ necessary for glycolysis. It is crucial in producing yogurt, sauerkraut, and during anaerobic exercise (Madigan et al. 1997).
2. Alcoholic Fermentation:
    - Yeasts and some bacteria convert glucose into ethanol and carbon dioxide, essential for brewing, winemaking, and bread making.
3. Mixed-Acid Fermentation:
    - Used by various enteric bacteria, this process produces a mixture of acids (lactic, acetic, succinic) and gases (CO_2, H_2). It plays a significant role in the human gut and industrial processes like solvent and biofuel production (Madigan et al. 1997).

1.3.3 Energy Generation in Anaerobes

Anaerobes generate energy through metabolic processes that do not involve oxygen, primarily through anaerobic respiration and fermentation, allowing them to thrive in oxygen-free environments.

1.3.3.1 Anaerobic Respiration

- This process uses alternative electron acceptors instead of oxygen, including nitrate (NO_3^-), sulphate (SO_4^{2-}), and carbon dioxide (CO_2). For example:
    - **Nitrate Respiration:** Nitrate is reduced to nitrogen gas (N_2) by denitrifying bacteria.
    - **Sulphate Respiration:** Sulphate is reduced to Hydrogen Sulphide (H_2S) by sulphate-reducing bacteria.
    - **Methanogenesis:** Methanogens reduce carbon dioxide to methane (CH_4) using hydrogen as an electron donor (Thauer et al. 2008).

The Electron Transport Chain (ETC) in anaerobic respiration transfers electrons from donors like NADH and $FADH_2$ to these alternative acceptors, generating a proton motive force that drives ATP synthesis via ATP synthase. Although anaerobic respiration yields less ATP than aerobic respiration, it is highly efficient under anoxic conditions.

1.3.3.2 Fermentation

- This process involves the partial oxidation of organic substrates, such as glucose, without an external electron acceptor. Key types of fermentation include:
  - **Lactic Acid Fermentation:** Converts glucose to lactic acid, regenerating NAD^+ for glycolysis.
  - **Alcoholic Fermentation:** Yeasts convert glucose to ethanol and CO_2, also regenerating NAD^+ (Madigan et al. 1997).

Fermentation generates ATP through substrate-level phosphorylation during glycolysis. While the ATP yield is lower compared to respiration, it allows anaerobes to survive and grow in oxygen-free environments.

1.4. Bacterial Anaerobes

1.4.1 Classification of Anaerobic Bacteria

Anaerobic bacteria thrive in environments devoid of oxygen and are classified based on their oxygen tolerance and metabolic activities. The primary groups include obligate anaerobes, which cannot survive in the presence of oxygen, and facultative anaerobes, which can grow with or without oxygen. They are also classified based on Gram staining into Gram-positive and Gram-negative bacteria.

Gram-Positive Anaerobes:

1. ***Clostridium* species:**
   - Notable for producing potent toxins, such as *C. botulinum* and *C. difficile*.
2. **Actinomyces:**
   - Filamentous bacteria causing *actinomycosis*.
3. **Propionibacterium species:**
   - Linked to acne and other skin conditions, like *P. acnes*.

Gram-Negative Anaerobes:

1. **Bacteroides:**
   - Particularly *B. fragilis*, significant in gastrointestinal infections.
2. **Fusobacterium species:**
   - Associated with periodontal diseases, such as *F. nucleatum*.
3. **Prevotella species:**
   - Involved in oral and respiratory infections, like *P. intermedia*.

Anaerobic bacteria play crucial roles in various ecosystems, including the human gut, aiding in digestion, and influencing immune responses. However, they can also be pathogenic, causing infections in low-oxygen tissues like deep wounds and abscesses. Effective treatment often involves specific antibiotics targeting anaerobic bacteria and surgical interventions to remove necrotic tissue (Zhang et al. 2015).

1.4.2 Ecological Niches of Anaerobic Bacteria

Anaerobic bacteria occupy diverse ecological niches where oxygen is limited or absent. These environments include the human gastrointestinal tract, soil, sediments, deep-sea vents, and various other natural and man-made anaerobic habitats. The metabolic versatility of anaerobic bacteria allows them to thrive in these environments, playing essential roles in biogeochemical cycles and contributing to the health and disease states of host organisms (Zhang and Okabe 2020).

1.4.2.1 Human Gastrointestinal Tract

The gut is a major niche for anaerobic bacteria, where they constitute the majority of the microbiota. Genera such as Bacteroides, *Clostridium*, and *Fusobacterium* dominate, aiding in the digestion of complex carbohydrates, synthesis of essential vitamins, and modulation of the host immune system. Their metabolic activities produce Short-Chain Fatty Acids (SCFAs) like acetate, propionate, and butyrate, crucial for colonic health and energy metabolism (Hillman et al. 2017).

1.4.2.2 Soil and Sediments

Anaerobic bacteria in soil and sediments, such as *Desulfovibrio* and *Methanobacterium*, play critical roles in the degradation of organic matter and cycling of sulphur and carbon. Sulphate-Reducing Bacteria (SRB) like Desulfovibrio contribute to the sulphur cycle by reducing sulphate to hydrogen sulphide, while methanogenic archaea like *Methanobacterium* participate in the carbon cycle by producing methane from carbon dioxide and hydrogen (Yuan et al. 2019).

1.4.2.3 Deep-Sea Vents and Extreme Environments

Anaerobic bacteria such as *Thermococcus* and *Pyrococcus* can create the Earth's extreme environmental living conditions by life that consumes hydrogen and sulphur compounds contributing to the unusual ecosystems existing around the hydrothermal vents (Poli et al. 2017). Besides, thermophilic and hyperthermophilic microorganisms utilize hydrogen and sulphur compounds as energy sources.

1.4.2.4 Bioreactors and Waste Treatment

A bioreactor is an anaerobic bacterial ecosystem that is found in fabrication and waste treatment, which is a process that is carried out by bacteria. Species like *Methanosaeta* and *Methanosarcina* are some of the most excellent bacteria and they are anaerobic bacteria. Using anaerobic digestion, they can convert organic waste into biogas (methane and carbon dioxide), which are alternative energy sources (Vítězová et al. 2020). This function is vital for managing waste and reducing environmental pollution. In conclusion, anaerobic bacteria are important agents in a variety of ecological niches and ecological niches. Their metabolic processes support the habitats they live in and have significant effects on industry, environmental sustainability, and human health.

1.4.2.5 Examples of Common Anaerobic Bacteria

Anaerobic bacteria are tiny organisms surviving and thriving in oxygen-deficient environments. They are crucial to several ecosystems and can even be pathogenic. Some of the most common examples are given as follows:

***Clostridium* species:**

- *Clostridium botulinum*: generates the botulinum toxin, which causes botulism.
- *Clostridium difficile*: causes colitis and severe diarrhoea, especially in hospital environments.
- *Clostridium perfringens*: Associated with food poisoning and gas gangrene.

***Bacteroides* species:**

- *Bacteroides fragilis*: A vital part of the gut microbiota, but when it becomes displaced, it can lead to infections inside the abdomen.
- *Bacteroides thetaiotaomicron*: Aids in the human gut's digestion of complex carbohydrates.

***Fusobacterium* species:**

- *Fusobacterium nucleatum*: Connected to periodontal disease and may play a role in the development of systemic infections such as Lemierre's syndrome.

Actinomyces species:

- *Actinomyces israelii*: Causes actinomycosis, a long-term infection that develops as oral, lung, or gastrointestinal ulcers.

Propionibacterium species:

- *Propionibacterium acnes*: frequently observed on skin and linked to the formation of acne.

Anaerobic bacteria live in humans and ecosystems and serve important functions but can also invade sterile body parts leading to severe infections.

1.5 Archaeal Methanogens

One of the unique archaeans known as methanogens produces methane during their metabolism when exposed to anaerobic conditions. They are the most significant participants in the carbon cycle and have been detected in different places, such as marshes, the digestive systems of ruminants, and anaerobic sediments (Jabłoński et al. 2015).

1.5.1 Characteristics of Methanogens

1. **Metabolism:** Various types of substrates are consumed by methanogens to produce methane, among which Carbon Oxide (CO_2) and Hydrogen (H_2) are the main but also formate, acetate, methanol, and methylamines. Methanogenesis, the central metabolic process, works through the reduction of CO_2 with H_2 to develop methane (CH_4) and water. This process is essential when there are no alternative electron acceptors present, such as oxygen, nitrate, sulphate, or iron.
2. **Cellular Structure:** Methanogens are a unique type of methanogens that are not similar to other forms of the bacteria archaea. They possess bacterial cell walls composed of pseudopeptidoglycan or other polymers, but not peptidoglycan. It provides stability under extreme conditions (Thauer et al. 2008).
3. **Environmental Adaptations:** Methanogens thrive in strictly anaerobic conditions. They are found in various habitats, including:
   - **Wetlands:** These are an important source of atmospheric methane.
   - **Digestive Tracts:** Especially found in ruminant animals, cows, sheep, etc., where they aid digestion.
   - **Sediments:** Freshwater sediments and marine environment.
   - **Extreme Environments:** Some methanogens, such as those in hydrothermal vents, can withstand high temperatures and pressures.
4. **Ecological Importance:** Methanogens play a crucial role in the global carbon cycle. In anaerobic environments, they are the final step in the decomposition of organic matter, converting complex organic compounds into methane. This process helps maintain the balance of carbon and other nutrients in ecosystems.
5. **Industrial and Environmental Applications:** Methanogens are utilized in biogas production. Biogas is a green-coloured organic waste gas that is changed to methane, a green and renewable energy source. They are also studied for their capability of reducing greenhouse chemical emissions, by managing methane production in agriculture.

1.5.2 Methanogenesis Pathways

Methanogenesis is the process by which methanogens, a group of archaea, produce methane (CH_4) under anaerobic conditions. This biochemical process is a key element of the carbon cycle, it occurs mainly in the anaerobic environment (Conrad 2005). Methanogenesis has three primary pathways: hydrogenotrophic, aceticlastic, and methylotrophic.

1. **Hydrogenotrophic Methanogenesis:**

   This pathway involves the reduction of carbon dioxide (CO_2) with hydrogen (H_2) as the electron donor:

   $$CO_2 + 4H_2 \rightarrow CH_4 + 2H_2O$$

   Key enzymes that help in this pathway are formylmethanofuran dehydrogenase, methylene-H4MPT dehydrogenase, and methyl-CoM reductase. Such a pathway is found in places where hydrogen is quite prevalent, for instance, the hydrothermal vents, and the stomachs of the ruminant animals.

2. **Aceticlastic Methanogenesis:**

   In this pathway, acetate (CH_3COOH) is cleaved to produce methane and carbon dioxide:

   $$CH_3COOH \rightarrow CH_4 + CO_2$$

   Acetate kinase is the main enzyme responsible for Aceticlastic methanogenesis. This pathway is significant in anaerobic environments rich in acetate, such as sediments and wetlands. *Methanosarcina* and *Methanosaeta* are the chief aceticlastic methanogens species involved in this process.

3. **Methylotrophic Methanogenesis:**

   This process is the production of methane from methylated compounds like methanol, methylamines, and methyl sulphides:

   $$4CH_3OH \rightarrow 3CH_4 + CO_2 + 2H_2O$$

   Methylotrophic methanogenesis is carried out by methanogens capable of utilizing these compounds, often found in environments like saline and hypersaline habitats. The key enzyme in this pathway is methyltransferase.

1.5.2.1 Ecological and Environmental Significance

Methanogenesis plays a vital role in the anaerobic breakdown of organic matter and the carbon cycle. Understanding these pathways aids in the development of mitigation methods for methane emissions from both natural and man-made sources. Methane produced through these channels is a potent greenhouse gas (Amin et al. 2021).

Applications:

Methanogenesis is a process used to produce biogas, which in turn uses organic waste for the production of methane, a renewable energy source. Understanding the methods used in methanation is quite important for treating methane emissions properly in waste management and agriculture.

1.6 The Role of Anaerobes in Human Health

Louis Pasteur's research on fermentation processes in the 19th century led to the identification of anaerobic microorganisms (Sebald and Hauser 1995). Manufacturers summoned Louis Pasteur to investigate the source of fermentation in preserved foods heated and sealed in containers after his crystallography and molecular asymmetry research revealed a connection between microorganisms and fermentation processes (Sebald and Hauser 1995). Pasteur dismissed Gya-lussac's suggestions regarding the Appert process and determined that facultative anaerobic yeasts were responsible for the fermentation of milk sugar. In 1861, he presented evidence that the Butyric

Vibrion, a microbe that thrives in an oxygen-free environment and exemplifies life without air, was responsible for butyrate fermentation (Sebald and Hauser 1995). In 1865, he additionally found that these vibrions had a role in the fermentation process of calcium tartrate and calcium lactate. He subsequently labels them as "anaerobic" microbes, incapable of reproducing in the presence of atmospheric oxygen (Ndongo et al. 2020). They play an essential role in biogas production, nitrogen cycling, and decomposition, among other ecological processes. In various habitats, you might encounter the following kinds of anaerobic microbes (Fig. 1.1).

Figure 1.2 describes the key ways in which anaerobic microbes influence human health, including physiological and pathological processes. Anaerobes are vital for preserving human health by contributing to important physiologic processes as well as having roles in various types of infections (Wang et al. 2017). To develop effective treatments and maintain a healthy microbiome, it is essential to acquire an overall understanding of the two functions they perform. With every new study, these bacteria become more intricate and treatments that change the face of healthcare will be developed as well as improve its outcomes.

In the complex organism, which is human beings, we can find bacterial and human cells that make up its cellular communities, tissues, and functioning organs. Humans as holobionts implicate the humans themselves together with their gut microbiota (Van De Guchte et al. 2018). The microbial communities that inhabit several ecological niches of the body such as the intestines are, depend on the human body (Van Der Ark et al. 2017). The human GI tract houses various microbial organisms like bacteria, fungi, archaea, viruses, etc. Most of the human's microbiota lives in the large intestine where facultative anaerobes dominate. More than 99% of wild-type species are unable to survive under oxygen conditions and so the majority of gut microorganisms in humans are anaerobes (Eckburg et al. 2005).

These anaerobes' behaviour and functions can be influenced by different circumstances such as the genetic makeup of the host, microbial colonization and succession after birth and use of antibiotics. The fact is becoming more apparent that nutrition significantly affects the composition and functionality of gut microbiota which in turn impacts on both gut health and general well-being (Johnson et al. 2019). Gut microbiota, containing indigenous microorganisms, plays key roles including nutrient acquisition, protection from pathogenic invaders and development/support for immune system functionality (Hegazy et al. 2017). A wide array of metabolites is produced by gut microbiota, some of which may have physiological importance to both gut and systemic health (Pope et al. 2017b, Fang et al. 2020). The composition as well as activity of microbiota in different physiological locations including but not limited to beyond the gastrointestinal tract and/or gut-brain axis can determine host health (David et al. 2013, Heijtz et al. 2011, Cryan and Mazmanian 2022).

Metagenomic technologies have brought attention to the profound impact these microbes have on human health and illness prevention, disruptions to their composition have long been linked to a wide range of diseases (Neef and Sanz 2013).

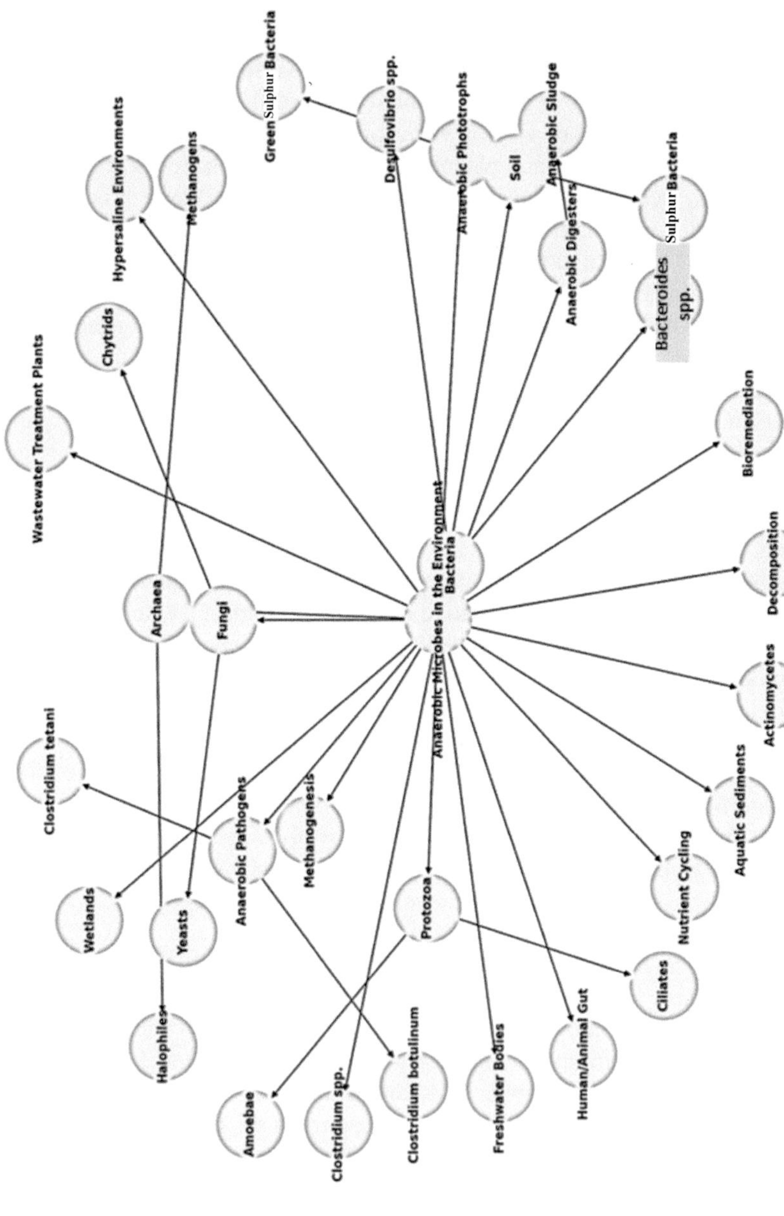

Figure 1.1. Types of Anaerobic microbes present in the environment.

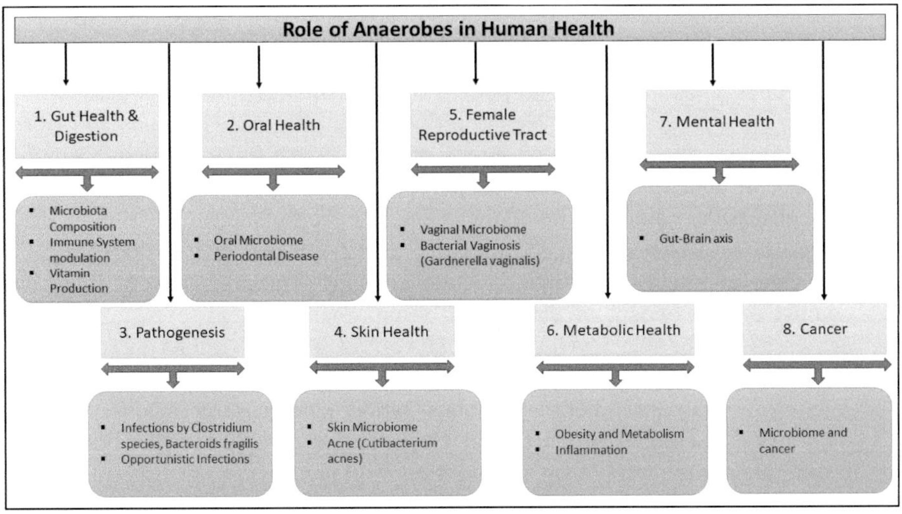

Figure 1.2. Role of anaerobes in Human Health.

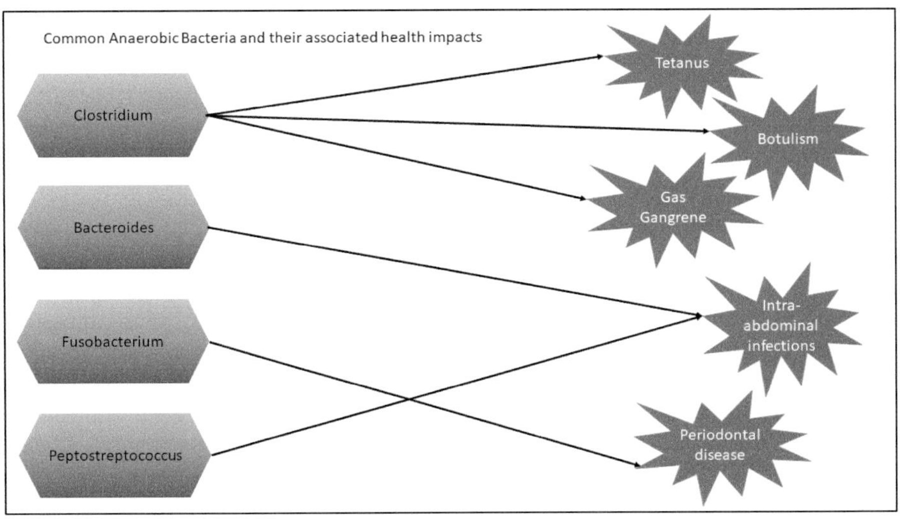

Figure 1.3. Common anaerobic bacteria and their associated health impacts.

Figure 1.3 shows the common anaerobic bacteria and the specific health conditions they are associated with.

1.6.1 Beneficial Roles of Anaerobic Microbes

The human gastrointestinal (GI) tract is an intricate and ever-changing ecology that harbours a large collection of bacteria, referred to as the gut microbiome. This community comprises bacteria, archaea, viruses, and fungi, with bacteria being the predominant group (Dekaboruah et al. 2020). Within this group of bacteria, anaerobes such as Bacteroides, *Clostridium*, and *Bifidobacterium* have significant importance in preserving the health of the gut and promoting overall well-being.

1. Anaerobes enhance the assimilation of nutrients that would otherwise be unattainable to the host by breaking down complex carbohydrates and fibres into simpler molecules. Short-Chain Fatty Acids (SCFAs) generated during the process of anaerobic fermentation play a crucial role in providing energy to the cells of the colon and maintaining overall energy balance. SCFAs, namely butyrate, aid in preserving the integrity of the gut barrier by stimulating the synthesis of mucins and tight junction proteins. These substances safeguard against the movement of detrimental microorganisms and toxins. Anaerobes are essential for the development and control of the immune system. They assist in regulating the immune system, fostering tolerance towards beneficial microorganisms and dietary antigens while safeguarding against harmful pathogens. Anaerobes produce vital vitamins that complement the host's diet, promoting overall well-being and averting shortages. Anaerobic bacteria engage in resource and attachment site competition with pathogenic microbes, generate antimicrobial chemicals, and regulate the gut environment to impede the growth of pathogens.

Bacteroides are Gram-negative bacteria that do not make spores and require an oxygen-free environment. They are highly skilled at decomposing complex carbohydrates, polysaccharides, and fibres that cannot be digested by human enzymes (Froidurot and Julliand 2022). This process entails the synthesis of diverse carbohydrate-active enzymes. Bacteroides generate Short-Chain Fatty Acids, including acetate, propionate, and butyrate, by fermenting food fibres (Fernández et al. 2016). SCFAs play a vital role in maintaining the health of the colon by acting as a source of energy for the cells that line the colon (colonocytes) and having anti-inflammatory qualities. Bacteroides have a role in the production of necessary vitamins, such as folate (vitamin B9) and biotin (vitamin B7), which are important for the body (Fernández et al. 2016).

Clostridium species are Gram-positive bacteria that form spores, thrive in the absence of oxygen, and metabolize food fibres and proteins to generate short-chain fatty acids (SCFAs) and gases (Rosenstrauch 2018). Butyrate, a type of short-chain fatty acid (SCFA) that is created in the gut, plays a crucial role in maintaining the health of the intestines. It serves as the main source of energy for the cells in the colon and helps to regulate the immune system (Silva et al. 2020). Certain species of *Clostridium* can produce necessary vitamins, including vitamin K and several B vitamins like riboflavin (vitamin B2) and niacin (vitamin B3). *Clostridium* species contribute to the formation and control of the immune system, aiding in the maintenance of immunological tolerance and the prevention of excessive immune reactions (Averianova et al. 2020).

Bifidobacterium species are Gram-positive bacteria that do not form spores and require an oxygen-free environment. They are highly skilled at breaking down oligosaccharides, especially those present in human milk, known as Human Milk Oligosaccharides or HMOs (Arzamasov et al. 2022). This skill is particularly crucial for newborns since it helps with their digestion and stimulates the development of helpful microorganisms (Saturio et al. 2021). Bifidobacteria generate short-chain fatty acids (SCFAs) via fermentation, which helps maintain gut health by reducing the pH of the gut environment and suppressing the growth of harmful bacteria. Bifidobacteria aid in the production of B vitamins, such as folate (vitamin B9) and cobalamin (vitamin B12), which are crucial for a range of metabolic functions.

1.6.2 Probiotics

According to the International Scientific Association for Probiotics and Prebiotics, probiotics are "live microorganisms that, when administered in adequate amounts, confer a health benefit on the host" (Hill et al. 2014). These microorganisms, primarily bacteria but also including yeasts, naturally occur in fermented foods, can be added to other food products, and are accessible as nutritional supplements. Nevertheless, it is important to note that not all food products and dietary supplements that are branded as probiotics in the market have been scientifically proven to provide health advantages.

Probiotics are distinguished by their unique strain, which consists of the genus, species, subspecies (if relevant), and an alphanumeric strain designation (WGO Review Team et al. 2017). The predominant microbial organisms utilized in probiotic products include *Lactobacillus*, *Bifidobacterium*, *Saccharomyces*, *Streptococcus*, *Enterococcus*, *Escherichia*, and *Bacillus*. Probiotic bacteria, whether consumed by food or supplements, are classified as part of our "transient microbiota" since their incorporation into the existing gut microbiota is temporary. It is crucial to recognize that the state of gut microbiota has a significant impact on the health and sickness of the host. Additionally, it is necessary to note that every individual has their resident gut microbiota (Jalanka-Tuovinen et al. 2011).

Probiotics can affect resident communities through three distinct mechanisms: (1) trophic interactions with resident members, (2) encouraging or inhibiting the growth of community members, and (3) eliciting a host reaction that indirectly alters the microbiota (Lobionda et al. 2019)

1.6.3 Immune System Modulation

When it comes to building and controlling the immune system, the anaerobic bacteria in our gut microbiome are crucial. To keep the immune response in check, prevent infections, and keep the gut barrier intact, they interact with the host's immune cells and the gut environment (Fig. 1.4).

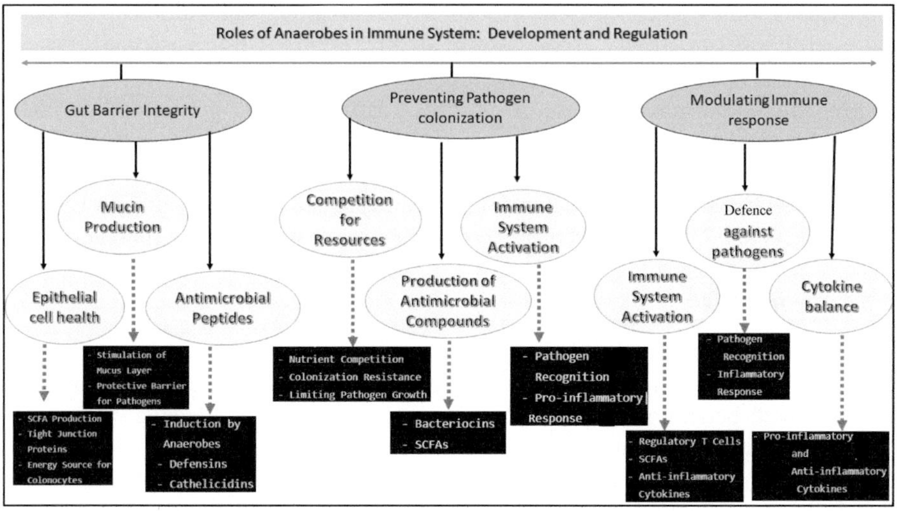

Figure 1.4. Role of anaerobes in regulating Immune System.

1.6.4 Maturation of the Immune System

The establishment of anaerobic bacteria such as Bacteroides, *Clostridium*, and Bifidobacterium in the gut starts at birth and plays a vital role in the development of the immune system. These bacteria promote the growth of Gut-Associated Lymphoid Tissue (GALT), which is an essential part of the immune system (Belkaid and Hand 2014). Anaerobes exert an influence on the process of immune cell differentiation, namely on T cells and regulatory T cells (Tregs). Tregs play a crucial role in preserving immunological tolerance and avoiding autoimmune responses (Oparaugo et al. 2023).

1.6.5 Regulation of Immune Responses

Pattern Recognition Receptors (PRRs) are molecules that are responsible for recognizing specific patterns in pathogens or other foreign substances. Anaerobes generate Microbial-Associated Molecular Patterns (MAMPs) that are detected by Pattern Recognition Receptors (PRRs) on immune cells. This connection facilitates the immune system's ability to differentiate between pathogenic microorganisms that cause harm and helpful microorganisms (Maier et al. 2014). Anti-inflammatory effects can be observed in certain anaerobes that produce metabolites such as short-chain fatty acids (SCFAs). Short-chain fatty acids (SCFAs), particularly butyrate, enhance the development of regulatory T cells (Tregs) and stimulate the production of anti-inflammatory cytokines such as IL-10 and IL-07. This leads to a reduction in inflammation and the promotion of immunological tolerance (Ney et al. 2023).

1.6.6 Preserving the Intestinal Barrier

Anaerobes promote the well-being of intestinal epithelial cells, which constitute the physical barrier of the gut (Gieryńska et al. 2022). SCFAs, namely butyrate, supply energy to these cells and enhance the production of tight junction proteins, which close the spaces between epithelial cells and hinder the escape of harmful microorganisms and poisons (Pérez-Reytor et al. 2021). Anaerobes promote the synthesis of mucins, which are glycoproteins that provide a defensive layer of mucus on the lining of the gastrointestinal tract. The mucus layer serves as a protective barrier against harmful microorganisms and provides a suitable environment for beneficial microorganisms (Corfield 2018). The gut epithelial cells can be stimulated by anaerobes to produce antimicrobial peptides (AMPs). Peptides like defensins and cathelicidins exhibit a wide range of antibacterial action and aid in regulating the proliferation of harmful microorganisms.

1.6.7 Pathogen Colonization Prevention and Resource Competition

Nutrient competition occurs when anaerobes outperform pathogenic bacteria in acquiring vital nutrients and securing attachment sites on the gut epithelium. This restricts the capacity of pathogens to establish themselves in the gut.

Colonization resistance refers to the ability of anaerobes to fill ecological niches and utilize available resources, hence creating an environment that is unfavourable for the growth and survival of diseases (Van Der Waaij and Nord 2000). Anaerobic microorganisms synthesize antimicrobial substances, including bacteriocins and SCFAs, that impede the proliferation of harmful bacteria. For instance, SCFAs decrease the pH of the gastrointestinal tract, establishing an inhospitable setting for numerous bacteria (Nogal et al. 2021). Anaerobes assist in triggering the immune system to initiate a defence when harmful bacteria penetrate the intestinal barrier. They enhance the generation of pro-inflammatory cytokines and attract immune cells such as neutrophils and macrophages to the infection site to eradicate the infections (Takiishi et al. 2017).

1.6.8 Regulating Immune Reactions

Immune tolerance is facilitated by regulatory T cells (Tregs), which are supported by anaerobes through the production of metabolites such as SCFAs (Kim 2023). These metabolites play a role in the differentiation and sustenance of Tregs. These cells play a critical role in preserving immunological tolerance to beneficial microorganisms and dietary antigens, hence reducing needless immune reactions and inflammation. Anaerobes contribute to the regulation of pro-inflammatory and anti-inflammatory cytokines, hence maintaining cytokine equilibrium. This equilibrium guarantees that the immune system can effectively react to infections without inducing excessive inflammation that may harm the host's tissues (Zhang and An 2007).

Anaerobes augment the immune system's capacity to detect and react to pathogens through pathogen recognition. They assist in priming immune cells, enhancing their ability to respond to pathogenic attacks. The presence of pathogens can trigger an orchestrated inflammatory response, which recruits immune cells to the site of infection and boosts the synthesis of antimicrobial peptides and other defence mechanisms (Mogensen 2009).

1.7 Ecological Significance

1.7.1 Anaerobes in Soil Ecosystems

Soil ecosystems are dynamic natural systems where living organisms significantly influence their physico-chemical characteristics. Soils comprise diverse habitats occupied by several organisms performing numerous functions (Waid 1999). Anaerobic bacteria are naturally found in compacted soils, deep within soil particles (microsites), and in hydric soils where the supply of oxygen is limited. Often pathogenic in nature, these bacteria can outcompete or even eliminate aerobic bacteria. Many anaerobes are also found in the intestines of animals and are also known to associate with manure and unpleasant odours (Lowenfels and Lewis 2006).

Most of the cultivated soil regions are dominated by aerobic conditions, however, there are certain circumstances where anaerobic conditions can also prevail. This in turn impacts soil health and plant productivity in multiple ways. The unfavourable effects of anaerobiosis on plants include root stress due to anoxia, nitrogen loss through denitrification, production of organic acids and H_2S, production of other plant toxicants, and conditions favouring the development of root pathogens. However, not all effects of anaerobic conditions are harmful (Tiedje et al. 1984). In the absence of oxygen, anaerobic microorganisms utilize alternative energy conservation strategies, enabling them to play vital roles in global cycles of carbon, nitrogen, and sulphur, and the degradation of recalcitrant compounds as well. These capabilities account for their potential usefulness in industrial and environmental biotechnology (Hatti-Kaul and Mattiasson 2016). Despite being less studied than aerobic microbes, anaerobes are now increasingly recognized and extensively studied for their diversity and metabolic potential.

1.7.1.1 Role in Soil Health

Soil enzymes play important roles in maintaining soil ecology and health. Most of these enzymes are of microbial origin and come from intracellular, cell-associated, or free enzymes. Consequently, microorganisms serve as indicators of soil health, significantly influencing nutrient cycling and impacting the soil's physico-chemical properties (Das and Varma 2011). Methanogenesis and methanotrophy are important phenomena taking place in mineral soils. Microbes producing methane are known as methanogens while the microbes which oxidize and consume methane are known as methanotrophs. These microorganisms work collectively to regulate methane flux balancing the processes of methanogenesis and methanotrophy (Reay et al. 2018). Anaerobic bacteria are also known to produce bioactive compounds, including antimicrobial agents, immunomodulators, and vitamins, contributing to soil and plant health (Mamo 2016).

1.7.1.2 Decomposition Process

The anaerobic decomposition of organic matter into methane and carbon dioxide is a complex microbial process that requires the syntrophic cooperation between anaerobic bacteria and methanogenic archaea. Fundamentally, biopolymers are hydrolyzed and fermented, with the resulting products being converted into compounds utilized by methanogens (Stams and Plugge 2010). This process includes hydrolysis, fermentation, and the production of Volatile Fatty Acids (VFAs), alcohols, and gases. Additionally, iron- and sulphate-reducing bacteria produce compounds like FeS along with the reduction of Fe^{3+} to Fe^{2+} and SO_4^{2-} to HS^- (Ueki et al. 2018). During the amino acid decomposition process, some anaerobes, like *Clostridium* spp., produce several phenolic compounds (Rainey et al. 2009). VFAs can lower soil pH, while nitrate reduction and ammonium production can increase it.

1.7.1.3 Pathogen Suppression

Biological Soil Disinfestation (BSD), also known as Reductive Soil Disinfestation (RSD), facilitates indigenous microbes to eliminate soil-borne plant pathogens. BSD treatments with organic materials control a wide range of pathogens effectively. In BSD-treated soil, anaerobic bacteria can produce enzymes with anti-pathogenic activities and release antagonistic substances against various soil pathogens (Larkin and Griffin 2007). BSD is an eco-friendly alternative to chemical fumigants. Employing numerous materials, research has shown that BSD suppresses a wide range of pathogens efficiently. These include *Agrobacterium tumefaciens*, *Fusarium oxysporum* f. sp. *spinaciae*, *Fusarium oxysporum* f. sp. *lycopersici*, *Pythium* spp., *Ralstonia solanacearum, Rhizoctonia solani, Sclerotium rolfsii, Verticillium dahliae*, as well as nematodes such as *Meloidogyne incognita* and *Pratylenchus* sp. (Butler et al. 2014, Muramoto et al. 2014, Ueki et al. 2018).

1.7.2 Anaerobic Environments in Aquatic Systems

Anaerobic environments in aquatic systems lack oxygen and are naturally found in areas with limited circulation and high organic matter, such as deep lake sediments and stagnant water columns. Such conditions are favourable for microorganisms that drive processes like sulphate reduction, methanogenesis, and denitrification (Diaz and Breitburg 2009). Methane production in ocean sediments is known to occur through both microbial and abiotic processes (Reay et al. 2018).

1.7.2.1 Biogeochemical Cycles

Although all living organisms contribute to biogeochemical cycles, the role of microorganisms is extremely important owing to their high abundance, vast metabolic capacities, and adaptability. These traits make them indispensable for the functioning and evolution of biogeochemical cycles. Consequently, microorganisms play a pivotal role in the adaptation, resistance, and resilience of ecosystems. Anaerobic bacteria in anoxic environments, such as sediments and wetland soils, play key roles in the biogeochemical cycles of elements like carbon, nitrogen, sulphur, and iron (Ueki et al. 2018).

Over 100 different free-living bacteria, both aerobic and anaerobic (as well as some actinomycetes and cyanobacteria), are utilized in converting nitrogen into ammonia. Examples include *Azotobacter* (aerobic), *Beijerinckia* (aerobic), *Azospirillum* (facultative), and *Clostridium* (anaerobic), all of which can fix nitrogen. A relatively recent discovery suggests that ammonium oxidation can also occur under anaerobic conditions using nitrite as the terminal electron acceptor (Kuenen 2008). This process is known as anammox which involves the bacterial oxidation of ammonium using nitrite, resulting in the production of nitrogen, which is critical for nitrogen cycling. Anammox bacteria thrive in environments where both ammonium and nitrite are present, typically found at aerobic-anaerobic interfaces. At these interfaces, ammonium is released from the degradation of organic matter (either aerobically or anaerobically) and from dissimilatory nitrate reduction to ammonium which is again an anaerobic process (Kuenen 2008).

The sulphur cycle involves the release of sulphur from organic forms, a process known as sulphur mineralization. This release occurs under both aerobic and anaerobic conditions. In the presence of oxygen, reduced sulphur compounds can support the growth of chemoautotrophic bacteria under strictly aerobic conditions and photoautotrophic bacteria under strictly anaerobic conditions. Sulphur and sulphate reducers, which are strict anaerobic chemoheterotrophic organisms, prefer small carbon substrates such as acetate, lactate, pyruvate, and low-molecular-weight alcohols (Maier 2015). In the case of the iron cycle, Fe^{3+} is reduced to the more soluble Fe^{2+} only under anaerobic conditions, facilitating its availability in the environment (Maier 2015).

1.7.3 Methane Production and Climate Change

Since the pre-industrial era, atmospheric methane concentrations have increased significantly, playing the role of a major contributor to global warming. Methane has a much higher global warming potential than carbon dioxide over 100 yrs. Microbial processes consume a significant portion of biogenic methane (Reeburgh 2007). Methanogens produce methane in anaerobic environments like anoxic water columns and freshwater lake sediments and utilize substrates produced by fermentative bacteria and eukaryotes (Hedderich and Whitman 2006). Methanogenesis accounts for a significant portion of carbon mineralization in these environments (Bastviken et al. 2008). Global warming increases methanogenesis by promoting archaeal growth. Furthermore, warming oceans destabilize methane hydrates, releasing methane and further contributing to climate change (Kurup and Kurup 2014).

1.8 Industrial Applications

1.8.1 Anaerobic Digestion for Biogas Production

Anaerobic digestion is a biochemical process in which microorganisms break down organic matter in the absence of oxygen and thus produce biogas which is mainly methane (CH_4) and carbon dioxide (CO_2) where the renewable energy source is derived (Deepanraj et al. 2014). This process is widely utilized for disposal and energy recovery, providing huge environmental and economic benefits.

Process Stages:

1. **Hydrolysis:** Complex organic materials such as carbohydrates, fats, and proteins are decomposed into simpler molecules like sugars, fatty acids, and amino acids.
2. **Acidogenesis:** These simpler molecules are further converted into Volatile Fatty Acids (VFAs), along with hydrogen and CO_2.
3. **Acetogenesis:** VFAs are transformed into acetic acid, hydrogen, and CO_2.
4. **Methanogenesis:** Methanogenic archaea convert acetic acid, hydrogen, and CO_2 into methane and water.

Feedstock and Applications:

Anaerobic digestion is a process that can deal with various organic waste types, such as agricultural leftovers, manure, municipal solid waste, industrial effluents, and sewage sludge. Biogas can be used for:

- **Electricity and Heat:** Through Combined Heat and Power (CHP) systems.
- **Vehicle Fuel:** As Compressed Natural Gas (CNG) after upgrading.
- **Pipeline Injection:** As a substitute for natural gas after purification.

Advantages:

- **Waste Reduction:** Effective management and reduction of organic waste volumes.
- **Renewable Energy:** Production of sustainable energy, reducing dependence on fossil fuels.
- **Nutrient Recycling:** Digestate, the byproduct of anaerobic digestion, can be utilized as a nutrient-rich fertilizer.

Anaerobic digestion of organic resources for biogas production is a versatile green method that reduces the problem of waste management and at the same time increases energy production and the sustainable management of the environment.

1.8.2 Anaerobic Fermentation in Food Industry

Anaerobic fermentation is the most important process in the food industry, as it allows the production of a wide variety of food and beverage products through the chemical reactions that happen inside the microorganisms in the absence of oxygen. This method serves the purpose of making foods more appetizing in terms of preservation, flavour, texture, and nutritional benefits (Anagnostopoulos and Tsaltas 2019).

Key Applications in the Food Industry:

1. **Dairy Products:**
   - **Yogurt and Cheese:** Lactose is fermented into lactic acid by certain lactic acid bacteria such as *Streptococcus* and *Lactobacillus* species. Yogurt is formed by an acidification process, which thickens the milk protein. While distinct flavour and texture are obtained by different fermentation conditions and different strains.

2. **Baked Goods:**
    - **Bread:** Yeasts, especially *Saccharomyces cerevisiae*, fermented sugars in the dough to produce carbon dioxide and ethanol. The CO_2 leavens the dough giving it that smooth spongy quality of bread while the ethanol evaporates on baking.
3. **Beverages:**
    - **Alcoholic Drinks:** The yeasts ferment sugars in grains, fruits, or other substrates to produce alcohol and CO_2. How the yeast becomes an antiseptic ethanol is fundamental to producing beer, wine, and spirits. Different yeast strains and growth conditions contribute to various flavour profiles.
4. **Fermented Vegetables:**
    - **Sauerkraut and Kimchi:** Lactic acid *Bacillus* ferments the sugars in cabbage and other vegetables, producing lactic acid. This acts as a natural preservative and gives the stuff its characteristic tangy flavour. This fermentation process also enhances the nutritional value by increasing the availability of vitamins and probiotics.
5. **Soy Products:**
    - **Tempeh and Miso:** Fermenting soybeans using fungi and bacteria substantially increases the protein content of products, and their digestibility, and gives them distinctive flavours. To make fermented tempeh, soybeans are fermented by *Rhizopus* species, to produce miso, soybeans and other grains are fermented by *Aspergillus oryzae* and lactic acid *Bacillus*.

Advantages of Anaerobic Fermentation:

- **Preservation:** The production of organic acids, alcohol, and other antimicrobial compounds inhibits spoilage organisms, extending the shelf life of fermented foods.
- **Flavour and Texture:** Fermentation enhances the sensory attributes of foods, creating complex flavours and desirable textures.
- **Nutritional Benefits:** Fermentation can increase the bioavailability of nutrients, introduce beneficial probiotics, and reduce anti-nutritional factors in food.

Anaerobic fermentation is a fundamental step of the food industry, which not only preserves but also makes a wide variety of healthy and tasty products. Whether fermented cream, bread, or kimchi, success with different kinds of food products is largely dependent on the microscopic choice and control of fermentation conditions.

1.8.3 Anaerobic Bioremediation

Anaerobic bioremediation using anaerobic microorganism is an environment friendly and efficient process to detoxify contaminated soil with pollutants. In this procedure, cells like these use the nutritional capacity of those species to eliminate or transform dangerous substances in places without oxygen. They may be found close to or below the currently available surface of the soil, in sediments used in dams or in the drinking water.

Mechanisms of Anaerobic Bioremediation:

1. **Reductive Dechlorination:** Anaerobic bacteria for example Dehalococcoides, are important players in chemical reduction which is the process of adding electrons to chlorinated organic pollutants like trichloroethylene (TCE) and perchloroethylene (PCE) sequentially. This is among those environmental rehabilitation procedures that are majorly used for contaminated industrial affirmatives (Löffler et al. 2013).
2. **Sulphate Reduction:** Sulphate-Reducing Bacteria (SRB) such as Desulfovibrio species convert sulphate to hydrogen sulphide (H_2S) while biodegrading organic contamination. Such a mechanism works very well for the treatment of heavy metal contamination in both ground and soil, as sulphide can cause precipitation of the metal and thus reduce the bioavailability and toxicity of the metal.
3. **Methanogenesis:** The archaea called methanogens break the organic matter down into methane and carbon dioxide during anaerobic digestion. Anaerobic digestion is the simplest and most common method that the solid and liquid organic-rich materials and landfills undergo to decay, where residues are biologically degraded using microorganisms, reducing the problem by one-third and producing biogas as a useful byproduct.
4. **Iron and Nitrate Reduction:** Some bacteria such as *Geobacter* species reduce ferric iron (Fe^{3+}) to ferrous iron (Fe^{2+}), while they also break down the organic matter. Correspondingly, nitrate-reducing bacteria are responsible for the conversion of nitrates to nitrogen gas, the process thus leading to the biodegradation of the contaminated groundwater by nitrates.

Applications and Benefits:

1. **Groundwater Remediation:** For most contaminated aquifers, it is applied anaerobic bioremediation, particularly in those places where it is hard to deliver oxygen. It is the best way to get rid of toxic metabolites such as chlorinated solvents, petroleum hydrocarbons, and heavy metals.
2. **Soil and Sediment treatment:** Anaerobic ways are used to treat contaminated soils and sediments, especially in marshes and beds of rivers, where oxygen is mostly not found naturally in water. This method of treatment allows us to bring back the ecosystem and decrease the danger caused the contaminants both human and animal at the same time.
3. **Industrial Waste Management:** Anaerobic bioremediation is one way through which industries try to clean wastewater and sludges, converting them to less harmful substances as well as simultaneously recovering energy in the form of biogas.

Anaerobic bioremediation can do all that. It helps clean up the environment and make it healthier. Because it uses the capacity of the microorganisms to metabolize anaerobes, it provides enough amount of energy, that is, it helps in reducing contamination and develops a healthy environment.

1.9 Challenges and Future Directions

The study of anaerobes and methanogens is crucial for understanding microbial life in oxygen-free environments. These organisms play vital roles in natural ecosystems and industrial processes. However, research in this field faces several challenges and presents promising future directions.

1.9.1 Challenges

1. **Cultivating difficulties:** A major number of anaerobes and methanogens are frustratingly not able to culture in the laboratory due to their strict environmental growth requirements, such as specific anaerobic conditions and complex nutrient needs (McGenity 2017). With this, the study of their physiology, metabolism, and genetic makeup in detail is only done to a certain extent.
2. **Environmental Complexity:** The success of anaerobes and methanogens in the most different and at the same time hardest conditions like in the deep-sea vents, even hot springs, or anoxic sediments. Reproducing these conditions in the laboratory for experimental studies is very hard and is usually performed using a combination of sophisticated equipment and techniques.
3. **Genomics and Metagenomics:** Though sequencing technology has advanced, genomics and metagenomics of the environment that lives without oxygen are very hard to analyze. It is difficult to fully understand the functional roles of these organisms in their entirety due to the presence of novel genes and metabolic pathways (Bharagava et al. 2019).
4. **Bioinformatics and the Data Integration:** One of the major bioinformatics challenges is to go through the large volume of data that was generated from genomic and metagenomic studies. It would also be a major but in reality, it should have been a major by combining this data with proteomic, transcriptomic, and metabolomic data to construct all-embracing models of bacterial communities that a long path is to cover.
5. **Impact of Environmental Changes:** Understanding how anaerobes and methanogens respond to changes in their environment, such as shifts in temperature, pH, and nutrient availability, is critical. Predicting their behaviour in response to climate change or anthropogenic impacts remains a significant challenge (Aronson et al. 2013).

1.9.2 Future Directions

1. **Advanced Cultivation Techniques:** Creating fresh and new cultivation methods like microfluidic systems and co-culturing techniques, can give the possibility to grow and study anaerobes and methanogens in controlled settings in a much faster and better way. This will give us a much deeper understanding of their physiology and functioning inside organisms.

2. **Synthetic and Systems Biology:** Implementing the engineering of synthetic biology pathways to make anaerobic and methanogenic cells create desired products is a way of smart technology. Systems biology can provide combined models to predict and impact the microbial community dynamics and metabolic networks.
3. **Environmental Biotechnology:** Anaerobes and methanogens can be used in bio-remediation and renewable energy production, which is seen in wastewater treatment and biogas generation, and serve sustainable forms to solve environmental challenges.
4. **Ecological and Evolutionary Studies:** Investigating the ecological roles and evolutionary adaptations of anaerobes and methanogens in natural environments can provide insights into their resilience and survival strategies. This knowledge is crucial for understanding global biogeochemical cycles.
5. **Climate Change Research:** Given the significant role of methanogens in methane production, studying their responses to environmental changes is essential for predicting and mitigating greenhouse gas emissions. This research can help to improve climate models and the decisions made regarding climate policy.
6. **Interdisciplinary Approaches:** Collaborations across microbiology, environmental science, engineering, and computational biology will be essential to address the complex questions surrounding anaerobes and methanogens. Interdisciplinary research can drive innovation and application in both basic and applied sciences.

Addressing these challenges and exploring future directions will enhance our understanding of anaerobes and methanogens, unlocking their potential in various scientific and industrial domains. The continued advancement in this field will contribute to sustainable solutions for environmental and energy-related issues.

References

Allwood, A. C., Walter, M. R., Kamber, B. S., Marshall, C. P. and Burch, I. W. (2006). Stromatolite reef from the Early Archaean era of Australia. Nature 441(7094): 714–718.

Amin, F. R., Khalid, H., El-Mashad, H. M., Chen, C., Liu, G. and Zhang, R. (2021). Functions of bacteria and archaea participating in the bioconversion of organic waste for methane production. Science of the Total Environment 763: 143007.

Anagnostopoulos, D. A. and Tsaltas, D. (2019). Fermented foods and beverages. pp. 257–291. *In*: Innovations in Traditional Foods. Woodhead Publishing.

Appels, L., Baeyens, J., Degrève, J. and Dewil, R. (2008). Principles and potential of the anaerobic digestion of waste-activated sludge. Progress in Energy and Combustion Science 34(6): 755–781.

Aronson, E. L., Allison, S. D. and Helliker, B. R. (2013). Environmental impacts on the diversity of methane-cycling microbes and their resultant function. Frontiers in Microbiology 4: 225.

Arzamasov, A. A., Nakajima, A., Sakanaka, M., Ojima, M. N., Katayama, T., Rodionov, D. A. et al. (2022). Human milk oligosaccharide utilization in intestinal bifidobacteria is governed by global transcriptional regulator NagR. MSystems 7(5). https://doi.org/10.1128/msystems.00343-22.

Averianova, L. A., Balabanova, L. A., Son, O. M., Podvolotskaya, A. B. and Tekutyeva, L. A. (2020). Production of Vitamin B2 (Riboflavin) by microorganisms: an overview. Frontiers in Bioengineering and Biotechnology 8. https://doi.org/10.3389/fbioe.2020.570828.

Bastviken, D., Cole, J. J., Pace, M. L. and Van de Bogert, M. C. (2008). Fates of methane from different lake habitats: Connecting whole-lake budgets and CH_4 emissions. Journal of Geophysical Research: Biogeosciences 113(G2).

Belkaid, Y. and Hand, T. W. (2014). Role of the microbiota in immunity and inflammation. Cell 157(1): 121–141. https://doi.org/10.1016/j.cell.2014.03.011.

Bharagava, R. N., Purchase, D., Saxena, G. and Mulla, S. I. (2019). Applications of metagenomics in microbial bioremediation of pollutants: from genomics to environmental cleanup. pp. 459–477. *In*: Microbial Diversity in the Genomic Era. Academic press.

Blankenship, R. E. (2010). Early evolution of photosynthesis. Plant Physiology 154(2): 434–438.

Blokhina, O., Virolainen, E. and Fagerstedt, K. V. (2003). Antioxidants, oxidative damage and oxygen deprivation stress: a review. Annals of Botany 91(2): 179–194.

Brook, I. (2007). Anaerobic Infections: Diagnosis and Management. CRC Press.

Butler, D. M., Kokalis-Burelle, N., Albano, J. P., McCollum, T. G., Muramoto, J., Shennan, C. et al. (2014). Anaerobic soil disinfestation (ASD) combined with soil solarization as a methyl bromide alternative: vegetable crop performance and soil nutrient dynamics. Plant Soil 378: 365–381.

Canfield, D. E. (2004). The evolution of the Earth surface sulfur reservoir. American Journal of Science 304(10): 839–861.

Canfield, D. E. (2005). The early history of atmospheric oxygen: homage to Robert M. Garrels. Annu. Rev. Earth Planet. Sci. 33(1): 1–36.

Change, I. P. O. C. (2007). Climate change 2007: The physical science basis. Agenda 6(07): 333.

Conrad, R. (2005). Quantification of methanogenic pathways using stable carbon isotopic signatures: a review and a proposal. Organic Geochemistry 36(5): 739–752.

Conrad, R. (2007). Microbial ecology of methanogens and methanotrophs. Advances in Agronomy 96: 1–63.

Corfield, A. P. (2018). The interaction of the gut microbiota with the mucus barrier in health and disease in human. Microorganisms 6(3): 78. https://doi.org/10.3390/microorganisms6030078.

Cryan, J. F. and Mazmanian, S. K. (2022). Microbiota–brain axis: Context and causality. Science 376(6596): 938–939. https://doi.org/10.1126/science.abo4442.

Das, S. K. and Varma, A. (2011). Role of enzymes in maintaining soil health. Soil Enzymology 25–42.

David, L. A., Maurice, C. F., Carmody, R. N., Gootenberg, D. B., Button, J. E., Wolfe, B. E. et al. (2013). Diet rapidly and reproducibly alters the human gut microbiome. Nature 505(7484): 559–563. https://doi.org/10.1038/nature12820.

Deepanraj, B., Sivasubramanian, V. and Jayaraj, S. (2014). Biogas generation through anaerobic digestion process-an overview. Research Journal of Chemistry and Environment 18(1): 5.

Dekaboruah, E., Suryavanshi, M. V., Chettri, D. and Verma, A. K. (2020). Human microbiome: an academic update on human body site specific surveillance and its possible role. Archives of Microbiology 202(8): 2147–2167. https://doi.org/10.1007/s00203-020-01931-x.

Diaz, R. J. and Breitburg, D. L. (2009). The hypoxic environment. pp. 1–23. *In*: Fish Physiology (Vol. 27). Academic Press.

Eckburg, P. B., Bik, E. M., Bernstein, C. N., Purdom, E., Dethlefsen, L., Sargent, M. et al. (2005). Diversity of the human intestinal microbial flora. Science 308(5728): 1635–1638. https://doi.org/10.1126/science.1110591.

ElMekawy, A., Diels, L., De Wever, H. and Pant, D. (2013). Valorization of cereal based biorefinery byproducts: reality and expectations. Environmental Science & Technology 47(16): 9014–9027.

Fang, P., Kazmi, S., Jameson, K. and Hsiao, E. (2020). The microbiome as a modifier of neurodegenerative disease risk. Cell Host & Microbe 28(2): 201–222. https://doi.org/10.1016/j.chom.2020.06.008.

Fernández, J., Redondo-Blanco, S., Gutiérrez-Del-Río, I., Miguélez, E. M. and Villar, C. J. (2016). Colon microbiota fermentation of dietary prebiotics towards short-chain fatty acids and their roles as anti-inflammatory and antitumour agents: A review. Journal of Functional Foods 25: 511–522. https://doi.org/10.1016/j.jff.2016.06.032.

Froidurot, A. and Julliand, V. (2022). Cellulolytic bacteria in the large intestine of mammals. Gut Microbes 14(1). https://doi.org/10.1080/19490976.2022.2031694.

Garcia, J. L., Patel, B. K. and Ollivier, B. (2000). Taxonomic, phylogenetic, and ecological diversity of methanogenic Archaea. Anaerobe 6(4): 205–226.

Gerardi, M. H. (2003). The Microbiology of Anaerobic Digesters. John Wiley & Sons.

Gieryńska, M., Szulc-Dąbrowska, L., Struzik, J., Mielcarska, M. B. and Gregorczyk-Zboroch, K. P. (2022b). Integrity of the intestinal barrier: the involvement of epithelial cells and microbiota—a mutual relationship. Animals 12(2): 145. https://doi.org/10.3390/ani12020145.

Hatti-Kaul, R. and Mattiasson, B. (2016). Anaerobes in Industrial-and Environmental Biotechnology (pp. 1–33). Springer International Publishing.

Hedderich, R., & Whitman, W. B. (2006). Physiology and biochemistry of the methane-producing Archaea. The prokaryotes, 2, 1050-1079.

Hegazy, A. N., West, N. R., Stubbington, M. J., Wendt, E., Suijker, K. I., Datsi, A. et al. (2017). Circulating and tissue-resident CD4+ T cells with reactivity to intestinal microbiota are abundant in healthy individuals and function is altered during inflammation. Gastroenterology 153(5): 1320–1337.e16.

Heijtz, R. D., Wang, S., Anuar, F., Qian, Y., Björkholm, B., Samuelsson, A. et al. (2011). Normal gut microbiota modulates brain development and behavior. Proceedings of the National Academy of Sciences of the United States of America 108(7): 3047–3052. https://doi.org/10.1073/pnas.1010529108.

Hill, C., Guarner, F., Reid, G., Gibson, G. R., Merenstein, D. J., Pot, B. et al. (2014). The international scientific association for probiotics and prebiotics consensus statement on the scope and appropriate use of the term probiotic. Nature Reviews. Gastroenterology & Hepatology 11(8): 506–514. https://doi.org/10.1038/nrgastro.2014.66.

Hillman, E. T., Lu, H., Yao, T. and Nakatsu, C. H. (2017). Microbial ecology along the gastrointestinal tract. Microbes and Environments 32(4): 300–313.

Jabłoński, S., Rodowicz, P. and Łukaszewicz, M. (2015). Methanogenic archaea database containing physiological and biochemical characteristics. International Journal of Systematic and Evolutionary Microbiology 65(Pt_4): 1360–1368.

Jalanka-Tuovinen, J., Salonen, A. H., Nikkilä, J., Immonen, O., Kekkonen, R., Lahti, L. et al. (2011). Intestinal microbiota in healthy adults: temporal analysis reveals individual and common core and relation to intestinal symptoms. PloS One 6(7): e23035. https://doi.org/10.1371/journal.pone.0023035.

Jetten, M. S., Wagner, M., Fuerst, J., Van Loosdrecht, M., Kuenen, G. and Strous, M. (2001). Microbiology and application of the anaerobic ammonium oxidation ('anammox') process. Current Opinion in Biotechnology 12(3): 283–288.

Johnson, A. J., Vangay, P., Al-Ghalith, G. A., Hillmann, B. M., Ward, T. L., Shields-Cutler, R. R. et al. (2019). Daily sampling reveals personalized diet-microbiome associations in humans. Cell Host & Microbe 25(6): 789–802.e5. https://doi.org/10.1016/j.chom.2019.05.005.

Kasting, J. F. (2005). Methane and climate during the Precambrian era. Precambrian Research 137(3-4): 119–129.

Kent, J. A. (ed.). (2013). Handbook of Industrial Chemistry and Biotechnology. Springer Science & Business Media.

Kim, C. H. (2023). Complex regulatory effects of gut microbial short-chain fatty acids on immune tolerance and autoimmunity. Cellular & Molecular Immunology/Cellular & Molecular Immunology 20(4): 341–350. https://doi.org/10.1038/s41423-023-00987-1.

Koga, Y. and Morii, H. (2007). Biosynthesis of ether-type polar lipids in archaea and evolutionary considerations. Microbiology and Molecular Biology Reviews 71(1): 97–120.

Kuenen, J. (2008). Anammox bacteria: from discovery to application. Nat. Rev. Microbiol. 6: 320–326. https://doi.org/10.1038/nrmicro1857.

Kurup, R. and Kurup, P. A. (2014). Climate change and global warming is produced by human endosymbiotic archaeal overgrowth and methanogenesis. Advances in Natural Science 7(4): 62–66.

Larkin, R. P. and Griffin, T. S. (2007). Control of soil-borne potato diseases using Brassica green manures. Crop Prot. 26: 1067–1077.

Lecomte, S. M., Achouak, W., Abrouk, D., Heulin, T., Nesme, X. and Haichar, F. E. Z. (2018). Diversifying anaerobic respiration strategies to compete in the rhizosphere. Frontiers in Environmental Science 6: 139.

Lobionda, S., Sittipo, P., Kwon, H. Y. and Lee, Y. K. (2019). The role of gut microbiota in intestinal inflammation with respect to diet and extrinsic stressors. Microorganisms 7(8): 271. https://doi.org/10.3390/microorganisms7080271.

Löffler, F. E., Ritalahti, K. M. and Zinder, S. H. (2013). Dehalococcoides and reductive dechlorination of chlorinated solvents. Bioaugmentation for Groundwater Remediation 39–88.

Louis, P., Hold, G. L. and Flint, H. J. (2014). The gut microbiota, bacterial metabolites and colorectal cancer. Nature Reviews Microbiology 12(10): 661–672.

Lowenfels, J. and Lewis, W. (2006). Teaming with Microbes: A Gardener's Guide to the Soil Food Web, Chapter 3: Bacteria, Timber Press, Portland, Oregon.

Lyons, T. W., Reinhard, C. T. and Planavsky, N. J. (2014). The rise of oxygen in Earth's early ocean and atmosphere. Nature 506(7488): 307–315.

Madigan, M. T., Martinko, J. M. and Parker, J. (1997). Brock Biology of Microorganisms (Vol. 11). Upper Saddle River, NJ: Prentice hall.

Maier, E., Anderson, R. and Roy, N. (2014). Understanding how commensal obligate anaerobic bacteria regulate immune functions in the large intestine. Nutrients 7(1): 45–73. https://doi.org/10.3390/nu7010045.

Maier, R. M. (2015). Biogeochemical cycling. pp. 339–373. *In*: Environmental Microbiology. Academic Press.

Mamo, G. (2016). Anaerobes as sources of bioactive compounds and health promoting tools. *In*: Hatti-Kaul, R., Mamo, G. and Mattiasson, B. (eds.). Anaerobes in Biotechnology. Advances in Biochemical Engineering/Biotechnology, vol 156. Springer, Cham. https://doi.org/10.1007/10_2016_6.

McGenity, T. J. (2017). Introduction to the isolation and cultivation of microbes involved in the hydrocarbon cycle. Hydrocarbon and Lipid Microbiology Protocols: Isolation and Cultivation, 1–25.

Mitsch, W. J. and Gosselink, J. G. (2015). Wetlands. John Wiley & Sons.

Mogensen, T. H. (2009). Pathogen recognition and inflammatory signaling in innate immune defenses. Clinical Microbiology Reviews 22(2): 240–273. https://doi.org/10.1128/cmr.00046-08.

Muramoto, J., Shennan, C., Baird, G., Zavatta, M., Koike, S. T., Bolda, M. P. et al. (2014) Optimizing anaerobic soil disinfestation for California strawberries. Acta Hortic 1044: 215–220.

Muyzer, G. and Stams, A. J. (2008). The ecology and biotechnology of sulphate-reducing bacteria. Nature Reviews Microbiology 6(6): 441–454.

Ndongo, S., Khelaifia, S., Lagier, J. C. and Raoult, D. (2020). From anaerobes to aerointolerant prokaryotes. Human Microbiome Journal 15: 100068. https://doi.org/10.1016/j.humic.2019.100068.

Neef, A. and Sanz, Y. (2013). Future for probiotic science in functional food and dietary supplement development. Current Opinion in Clinical Nutrition and Metabolic Care 16(6): 679–687. https://doi.org/10.1097/mco.0b013e328365c258.

Ney, L. M., Wipplinger, M., Grossmann, M., Engert, N., Wegner, V. D. and Mosig, A. S. (2023). Short chain fatty acids: key regulators of the local and systemic immune response in inflammatory diseases and infections. Open Biology 13(3). https://doi.org/10.1098/rsob.230014.

Nogal, A., Valdes, A. M. and Menni, C. (2021). The role of short-chain fatty acids in the interplay between gut microbiota and diet in cardio-metabolic health. Gut Microbes 13(1). https://doi.org/10.1080/19490976.2021.1897212.

Oparaugo, N. C., Ouyang, K., Nguyen, N. P. N., Nelson, A. M. and Agak, G. W. (2023). Human regulatory T cells: understanding the role of Tregs in select autoimmune skin diseases and post-transplant nonmelanoma skin cancers. International Journal of Molecular Sciences 24(2): 1527. https://doi.org/10.3390/ijms24021527.

Pérez-Reytor, D. C., Puebla, C., Karahanian, E. and García, K. P. (2021). Use of short-chain fatty acids for the recovery of the intestinal epithelial barrier affected by bacterial toxins. Frontiers in Physiology 12. https://doi.org/10.3389/fphys.2021.650313.

Poli, A., Finore, I., Romano, I., Gioiello, A., Lama, L. and Nicolaus, B. (2017). Microbial diversity in extreme marine habitats and their biomolecules. Microorganisms 5(2): 25.

Pope, J. L., Tomkovich, S., Yang, Y. and Jobin, C. (2017b). Microbiota as a mediator of cancer progression and therapy. Translational Research 179: 139–154. https://doi.org/10.1016/j.trsl.2016.07.021.

Rainey, F. A., Hollen, B. J. and Small, A. (2009). Genus I. *Clostridium* Prazmowski 1880, 23AL. pp. 736–828. *In*: Whitman, W. B. and Parte, A. C. (eds.). Bergey's Manual of Systematic Bacteriology, vol 3. Springer, New York.

Reay, D. S., Smith, P., Christensen, T. R., James, R. H. and Clark, H. (2018). Methane and global environmental change. Annual Review of Environment and Resources 43: 165–192.

Reeburgh, W. S. (2007). Oceanic methane biogeochemistry. Chemical Reviews 107(2): 486–513.

Rosenstrauch, J. M. (2018). Impact of Bacillus probiotics on production performance and microbial gut diversity in weaner pigs. *In*: MSc (Agric) Animal Science: Animal Nutrition.

Saturio, S., Nogacka, A. M., Alvarado-Jasso, G. M., Salazar, N., De Los Reyes-Gavilán, C. G., Gueimonde, M. et al. (2021). Role of bifidobacteria on infant health. Microorganisms 9(12): 2415. https://doi.org/10.3390/microorganisms9122415.

Schink, B. (1997). Energetics of syntrophic cooperation in methanogenic degradation. Microbiology and Molecular Biology Reviews 61(2): 262–280.

Schmidt, O., Hink, L., Horn, M. A. and Drake, H. L. (2016). Peat: home to novel syntrophic species that feed acetate-and hydrogen-scavenging methanogens. The ISME Journal 10(8): 1954–1966.

Sebald, M. and Hauser, D. (1995). Pasteur, oxygen and the anaerobes revisited. Anaerobe 1(1): 11–16. https://doi.org/10.1016/s1075-9964(95)80353-x.

Silva, Y. P., Bernardi, A. and Frozza, R. L. (2020). The role of short-chain fatty acids from gut microbiota in gut-brain communication. Frontiers in Endocrinology 11. https://doi.org/10.3389/fendo.2020.00025.

Stams, A. J. and Plugge, C. M. (2010). The microbiology of methanogenesis. pp. 14–26. *In*: Methane and Climate Change. Routledge.

Takiishi, T., Fenero, C. I. M. and Câmara, N. O. S. (2017). Intestinal barrier and gut microbiota: Shaping our immune responses throughout life. Tissue Barriers 5(4): e1373208. https://doi.org/10.1080/21688370.2017.1373208.

Thauer, R. K., Kaster, A. K., Seedorf, H., Buckel, W. and Hedderich, R. (2008). Methanogenic archaea: ecologically relevant differences in energy conservation. Nature Reviews Microbiology 6(8): 579–591.

Tiedje, J. M., Sexstone, A. J., Parkin, T. B. and Revsbech, N. P. (1984). Anaerobic processes in soil. Plant and Soil 76: 197–212.

Todar, K. (2006). Todar's Online Textbook of Bacteriology.

Ueki, A., Kaku, N. and Ueki, K. (2018). Role of anaerobic bacteria in biological soil disinfestation for elimination of soil-borne plant pathogens in agriculture. Appl. Microbiol. Biotechnol. 102: 6309–6318. https://doi.org/10.1007/s00253-018-9119-x.

Ueno, Y., Yamada, K., Yoshida, N., Maruyama, S. and Isozaki, Y. (2006). Evidence from fluid inclusions for microbial methanogenesis in the early Archaean era. Nature 440(7083): 516–519.

Van De Guchte, M., Blottière, H. M. and Doré, J. (2018). Humans as holobionts: implications for prevention and therapy. Microbiome 6(1). https://doi.org/10.1186/s40168-018-0466-8.

Van de Werken, H. J. G. (2008). Computational Genomics of Hyperthermophiles (Publication No. 28238307) [Doctoral dissertation, Wageningen University and Research]. ProQuest Dissertations & Theses.

Van Der Ark, K. C., Nugroho, A. D. W., Berton-Carabin, C., Wang, C., Belzer, C., De Vos, W. M. et al. (2017). Encapsulation of the therapeutic microbe Akkermansia muciniphila in a double emulsion enhances survival in simulated gastric conditions. Food Research International 102: 372–379. https://doi.org/10.1016/j.foodres.2017.09.004.

Van Der Waaij, D. and Nord, C. E. (2000). Development and persistence of multi-resistance to antibiotics in bacteria, an analysis and a new approach to this urgent problem. International Journal of Antimicrobial Agents 16(3): 191–197. https://doi.org/10.1016/s0924-8579(00)00227-2.

Vítězová, M., Kohoutová, A., Vítěz, T., Hanišáková, N. and Kushkevych, I. (2020). Methanogenic microorganisms in industrial wastewater anaerobic treatment. Processes 8(12): 1546.

Waid, J. S. (1999). Does soil biodiversity depend upon metabiotic activity and influences? Applied Soil Ecology 13(2): 151–158.

Wang, B., Yao, M., Lv, L., Ling, Z. and Li, L. (2017). The human microbiota in health and disease. Engineering 3(1): 71–82. https://doi.org/10.1016/j.eng.2017.01.008.

WGO Review Team, Guarner, F., Sanders, M. E., Eliakim, R., Fedorak, R., Gangl, A. et al. (2017). World Gastroenterology Organisation Global Guidelines Probiotics and Prebiotics.

Yuan, C. L., Zhang, L. M., Wang, J. T., Hu, H. W., Shen, J. P., Cao, P. et al. (2019). Distributions and environmental drivers of archaea and bacteria in paddy soils. Journal of Soils and Sediments 19: 23–37.

Zhang, J. M. and An, J. (2007). Cytokines, inflammation, and pain. International Anesthesiology Clinics 45(2): 27–37. https://doi.org/10.1097/aia.0b013e318034194e.

Zhang, L. and Okabe, S. (2020). Ecological niche differentiation among anammox bacteria. Water Research 171: 115468.

Zhang, Y. J., Li, S., Gan, R. Y., Zhou, T., Xu, D. P. and Li, H. B. (2015). Impacts of gut bacteria on human health and diseases. International Journal of Molecular Sciences 16(4): 7493–7519.

Chapter 2

Adapting to Anaerobiosis

Physiology and Molecular Responses

Chamma Gupta,[1] *Bidita Khandelwal,*[2,*] *Abhishek Byahut*[3] and *Vaishnavi Singh*[4]

2.1 Introduction

The ability of living organisms to adapt to anaerobic conditions, characterized by a scarcity of oxygen, underscores the remarkable diversity and resilience of life on Earth. Human endeavours are changing the climate worldwide, altering the ecological interactions of invasive species, and creating pollution in the ecosystems. Multidisciplinary biological science is at the forefront of our understanding of how changes in the global environment affect the diversity and bioactivity of species (Todgham and Stillman 2013). There are a multitude of distinct species of anaerobic bacteria that can thrive in conditions with extremities of temperature, pH levels, salt content, toxicity of the substrate, ambient energies and/or in limited supply of oxygen. One of the primary elements influencing the growth of bacteria is oxygen. Anaerobes are microbes that can survive and function in oxygen-free or extremely low-oxygen conditions, as opposed to aerobes, which require oxygen for metabolic functions. Obligate, facultative, aerotolerant anaerobes, and microaerophiles are the

[1] Lecturer, Department of Medical Biotechnology, Sikkim Manipal Institute of Medical Sciences, Sikkim Manipal University (SMU), Gangtok, Sikkim, India, 737102.
[2] Professor and HOD, Department of Medicine, Sikkim Manipal Institute of Medical Sciences, Sikkim Manipal University (SMU), Gangtok, Sikkim, India, 737102.
[3] PhD Scholar, Department of Medical Biotechnology, Sikkim Manipal Institute of Medical Sciences, Sikkim Manipal University (SMU), Gangtok, Sikkim, India, 737102.
[4] MSc Med. Biotechnology, Department of Medical Biotechnology, Sikkim Manipal Institute of Medical Sciences, Sikkim Manipal University (SMU), Gangtok, Sikkim, India, 737102.
Emails: chamma28feb@gmail.com, abhishekbyahut93@gmail.com, singhvaishnavi1212@gmail.com
* Corresponding author: drbidita@gmail.com

DOI: 10.1201/9781003506898-2

varied categories of anaerobes. While facultative anaerobes can alternate between aerobic and anaerobic metabolism, microaerophiles need low oxygen concentrations, aerotolerant anaerobes can tolerate its presence, while obligatory anaerobes are exclusively anaerobic and lethal to oxygen (Hentges 1996).

The capacity to flourish in anaerobic settings is one of the important adaptive mechanisms that organisms have acquired to deal with a variety of situations. Survival under anaerobiosis, which needs intricate navigation and adaptation to these oxygen-deprived settings requires significant physiological and biochemical changes. It is thus crucial to comprehend the state on how various living forms, from microbes to higher organisms, traverse and adjust to these environments devoid of oxygen. An intricate web of molecular interactions, such as the control of gene expression, the initiation of stress responses, and the synthesis of specialized enzymes, are involved in an organism's adaptation to anaerobiosis. Gene expression patterns significantly alter in cells in response to oxygen deprivation. One typical reaction is the upregulation of genes linked to anaerobic metabolism, such as those implicated in pathways for fermentation or alternate respiratory processes. On the other hand, in an effort to save energy and resources, genes linked to aerobic activities could be downregulated. Synthesis of stress proteins and activation of various signalling pathways under a stress condition also paves a way to enhance cellular activities under oxygen-deficient conditions. These adaptive measures serve to enhance cell survival, maintain cellular homeostasis, and ensure the continuity of essential physiological processes. The oxygen content of a biotope governs its suitability for obligatory aerobes or anaerobes. Facultative anaerobic bacteria like *E. coli* may thrive in varying oxygen levels, by extensively adapting and altering their metabolic processes (Allen and Martin 2007).

In contrast to aerobic organisms, anaerobic bacteria seem to have inherited more energy-conserving physiological adaption strategies for ambient pressures, such as unique functional and structural confirmation of the lipid membrane, and distinct enzymatic activity and stability. Thermoanaerobes and associated stable enzymatic compounds have been used as biological model system for biomedical, fundamental, and therapeutic research toward the metabolism of microbes. These metabolic processes may include "cellulose as well as starch degradation, methanogenesis, ethanologenesis, acetogenesis, autotrophic CO_2 fixation, saccharidases, hydrogenases, and alcohol dehydrogenases". Aerobic bacterial equivalents are not comparable to anaerobic syntrophs. It is stated that the metabolic end product of the syntrophs requires synchronized utilization through a subsequent partner organism for the survival and growth of an organism (Zhu et al. 2020, Lowe et al. 1993).

The presence of oxygen indicates potential infection sites, as many areas within recipient systems are hypoxic or anoxic. When bacteria form biofilms during long-term infections, oxygen levels in the biofilm gradually falls until it reaches "zero dissolved oxygen (DO)" with a level of certain microns below the surface. However, reports suggest that finding a DO of zero doesn't always indicate anaerobic state, but instead that most of the oxygen that was moved to the liquefied state was utilized for breathing (Pedraz et al. 2020, Alexeeva et al. 2002). Anaerobes with a high temperature tolerance appear to have been predecessors of modern archaebacterium and eubacterium species. Psychrophilic microbes are highly adaptable to extreme frigid environmental conditions owing to their extensive metabolic alterations, which include altered food transport, intracellular ice production, and cold denaturation of proteins (Kumari et al. 2021). In addition to the physiological and molecular responses, anaerobiosis leads to morphological and structural changes in organisms, such as specialized cell membranes or altered cell membrane compositions, which are crucial for survival in oxygen-deprived environments. Understanding this interplay can aid in environmental studies, bioremediation, and biotechnology. The objective of the chapter is to present a thorough review of the physiological and molecular responses of an anaerobe to anaerobiosis, shedding light on the various mechanism and structural alterations used by an organism to survive in stringent conditions.

2.2 Physiological Adaptation

Organisms from various domains of life have developed unique strategies to adapt to anaerobiosis. From unicellular microorganisms to complex multicellular organisms, the ability to survive and thrive in the absence of oxygen involves a myriad of physiological and molecular adaptations (Cooper 2000).

2.2.1 Prokaryotic Organisms

In prokaryotic organisms such as bacteria and archaea, the transition to anaerobiosis involves the activation of specific metabolic pathways, including the utilization of alternative electron acceptors and the production of fermentation by products. These organisms have evolved intricate regulatory networks that govern the expression of genes involved in anaerobic metabolism, allowing them to efficiently generate energy in oxygen-deprived environments (Cole 2016).

2.2.2 Eukaryotic Organisms

In eukaryotic organisms, including yeast and certain parasitic organisms, the shift to anaerobiosis entails the reprogramming of cellular processes to sustain energy production. Eukaryotic cells undergo metabolic reorganization, accompanied by the upregulation of genomic transcripts that codes for enzymes functional in anaerobic metabolism and the downregulation of aerobic respiration-related genes. This adaptation enables these organisms to maintain essential cellular functions despite the lack of oxygen (Cole 2016).

2.3 Key Models Organism for Anaerobiosis

Common anaerobic model species like *Escherichia coli*, *Bacillus subtilis*, and *Pseudomonas aeruginosa* adapt to harsh environments through adaptation. The physiological response to anaerobiosis may occur via processes like respiration, fermentation, and denitrification.

2.3.1 *Escherichia coli*

Escherichia coli (*E. coli*), a facultative anaerobe, is commonly seen as an intestinal pathogen within the colon. These microorganisms can initiate infections internally as well as externally. Although the gut is thought to historically possess anaerobic conditions, it is now understood that the gut harbours substantial micro-aerobiosis owing to diffusion from vascularized tissue (Jones et al. 2007, Tenaillon et al. 2010, Hentges 1996). *E. coli's* primary mode of growing is aerobic respiration, which presents two primary alternate terminal oxidases: cytochrome bd-I, which has an extraordinarily high affinity for oxygen and may keep growing even at the lowest oxygen pressures, and cytochrome bo', which is mostly employed for highly aerated settings; cytochrome bd-II is a another, less explored oxidase. *E. coli* may proliferate by combined-acid fermentation or anaerobic respiration in the dearth of oxygen (Bettenbrock et al. 2014).

2.3.2 *Bacillus subtilis*

Bacillus subtilis, a Gram-positive soil bacterium, is commonly referred to as an aerobe, however, flourishes in anaerobic settings. There are two main methods that *Bacillus subtilis* grows anaerobically: fermentation when there are no electron acceptors and respiration when nitrate is used as an electron acceptor. In regard to variations in the presence of oxygen, transition has taken place regulating and enabling cells to switch to these physiological processes. Among these is the transcriptional activation of Fumarate and Nitrate Reduction (FNR) regulator in response to oxygen limitation, which is dependent on the two-component signal transduction pathway ResD-ResE, which is also involved in controlling aerobic respiration. Subsequently, FNR triggers the transcription of additional anaerobically stimulated genes, such as the respiratory nitrate reductase-encoding NarL-dependent upregulation of the nitrate reductase (narGHJI) operon. Unidentified FNR-independent regulatory mechanism controls genes that regulate fermentative activity (Nakano and Hulett 1997, Pieper et al. 2023).

2.3.3 *Pseudomonas aeruginosa*

Pseudomonas aeruginosa is a gram-negative bacterium residing in the soil, water and habitats related to humans, animals, and plant hosts and can thrive with low oxygen levels. *Pseudomonas aeruginosa* has a completely distinct response to oxygen constraints, even though it may be grown in both aerobic and anaerobic condition. *P. aeruginosa* are opportunistic pathogens associated with infectious conditions in people with inherent immune deficiencies like Cystic fibrosis (CF) or Chronic Obstructive Pulmonary Disease (COPD), a lung disease in which *P. aeruginosa* develops characteristic biofilms. Mucoid biofilms in CF tend to be oxygen-depleted (Sabra et al. 2002, Worlitzsch et al. 2002). Development and growth of *P. aeruginosa* can be brought on anaerobically by when l-arginine is used as a medium for proliferation or during existence of terminal electron acceptors such nitrate (NO_3^-), nitrite (NO_2), and nitrous oxide (N_2O) (Ghasemi 2022). Additionally, *P. aeruginosa* possesses the entire complement of denitrification enzymes, that uses nitrite,

Nitric Oxide (NO), and nitrous oxide to form nitrogen molecules (N) from nitrate. These NOs allow *P. aeruginosa* to flourish in anaerobic environments by acting as substitute electron acceptors. It is also anticipated that NO reductase, among the denitrification enzymatic agents, will aid in the detoxification of NO generated by the host immunological defence system. *P. aeruginosa* can adapt to a variety of ambient circumstances, including those seen in affected hosts, by controlling the production of various aerobic and anaerobic respiratory enzymes (Arai 2011).

2.4 Physiological Responses Anaerobiosis

2.4.1 Metabolic Shift and Fermentation

One of the primary cellular adaptations to oxygen limitation is the metabolic shift from aerobic respiration to anaerobic respiration or fermentation. This shift enables cells to continue generating ATP without any presence of oxygen. In anaerobic respiration, organisms utilize alternative electron acceptors such as nitrate, sulphate, or carbon dioxide to sustain their energy production. Similarly, fermentation allows for the generation of ATP through the partial oxidation of organic compounds, producing various end products such as ethanol, lactate, or other organic acids. Another crucial aspect of cellular adaptation to anaerobic state is the maintenance of redox balance. In the microenvironment lacking oxygen, cells must efficiently manage their redox state to prevent the accumulation of toxic by-products and ensure the proper functioning of essential metabolic reactions. This often involves the redistribution of reducing equivalents and the activation of specialized enzymes and metabolic pathways to maintain redox homeostasis (Clark 1989, Jurtshuk 1996).

Due to its facultative anaerobic nature, *E. coli* can survive both aerobically and anaerobically when carbohydrates like glucose are its only supply of carbon and energy. In the event of a lack of substitute electron acceptors and in an environment with scarce level of oxygen, fermentation of carbohydrates takes place to produce a variety of metabolites in *Escherichia coli*. Acetate, ethanol, lactate, and formate are the main soluble bio products, with reduced quantities of succinate. Additionally, substantial volumes of the gases such as carbon dioxide and hydrogen are released as by products. The fermented output pathways are branching and extended, and the dissociations occurs at each branch varying according to the fermentation substrate's characteristics as well as the culture medium's pH. To balance the quantity of reducing equivalents produced during the substrate's glycolytic breakdown, the ratios of the different fermentation products are specifically adjusted (Clark 1989).

Energy must be provided in the dearth of oxygen by fermentation or anaerobic respiration that utilizes electron acceptors such fumarate, nitrate, or trimethylamine oxide. The phosphotransferase system transports glucose into the cell regardless of oxygen levels, where it is metabolized to produce pyruvate (Clark 1989, Jurtshuk 1996). The variations between aerobic and anaerobic environments manifest themselves in the subsequent metabolism of pyruvate. In air, the respiratory chain's activity re-oxidizes NADH produced by the Krebs cycle, glycolysis, and related reactions. During microbial fermentation, neither the respiratory chains involving oxygen nor those associated to alternate electron acceptors are active.

Under anaerobic conditions, the large-scale NADH-generating Krebs cycle and pyruvate dehydrogenase processes halt working. The glycolytic sequence cannot, however, continue until the NADH generated by glycolysis is re-oxidized to NAD^+. Therefore, the primary concern in fermentation is the recycling of NADH through the fermentation product's conversion from pyruvate (Bragg 1980, Donnelly et al. 1998).

In most species, the transition from mitochondrial respiration to fermentation is the defining feature of anaerobic metabolism. This entails fermentation, glycolysis, and a changed metabolism of carbohydrates. In the absence of an electron acceptor, certain bacteria, such *E. coli*, ferment. In fermentation, electron transport processes are unable to re-oxidize the NADH liberated during the process of glycolysis. In contrast to respiration, which uses proton potential to synthesise ATP, NAD^+ is instead synthesized utilizing endogenous electron acceptors generated during the breakdown of pyruvate, whereas ATP is released by phosphorylation. Clark (Clark 1989) assessed fermentation taking place in *E. coli* and illuminated it from the perspective of redox balance. It is unclear how *B. subtilis* ferments some substances.

When *B. subtilis* ferments glucose, it does so poorly or not at all. However, *B. subtilis* cells proliferate by fermentation when exposed to glucose and pyruvate, or when they are exposed to glucose and a combination of twenty amino acids. The mechanism behind glucose fermentation promoted by pyruvate and inability of *B. subtilis* (unlike *E. coli*) to ferment either glucose or pyruvate effectively is not clearly understood. However, in anaerobic settings, *B. subtilis* was shown to produce most or all of its acetyl coenzyme A by pyruvate dehydrogenase (PDH), as opposed to *Escherichia coli*, which employs pyruvate formatelyase. Acetate, acetoin, ethanol, lactate, succinate, and 2,3-butanediol are among the fermentation products in *B. subtilis* as revealed in Nuclear Magnetic Resonance (NMR) spectroscopy analysis, suggesting a mixed acid fermentation (Nakano and Hulett 1997, Nakano et al. 1997, Schirawski and Unden 1995).

2.4.2 Nitrate Respiration

Nitrate serves as an important component in the assimilatory as well as dissimilatory pathways in microorganisms. Since nitrate (NO_3^-) has the second-best reduction potential after O_2, it is used as an alternate electron acceptor in place of O_2. During denitrification, reduction of nitrate take place through nitrite, molecular nitrogen (N_2) production through NO and N_2O in the presence of four membrane localized reductases with the synchronized generation of an ion gradient for ATP production (Acuna et al. 2017). Different enzymes are involved in nitrate respiration depending on the organism, converting NO_3^- to either dinitrogen (N_2) or ammonium (NH_4^+). Nar and Nap are the two important nitrate reductases involved in the nitrate respiration process, where Nar is a transmembrane protein receiving electron from quinone pool and enable Nar reductase to translocate the proton across the membrane, while the Nap reductase which is located in the periplasm acts as a membrane-bound quinol dehydrogenase as well as periplasmic electron transfer adapter protein to initiate the denitrification process (Durand and Guillier 2021).

A preliminary physiological investigation indicating that *B. subtilis* might thrive in the absence of oxygen revealed that altered membrane-bound nitrate reductase existed amid minimal culture oxygenation (Pierre et al. 1976), despite taxonomists using nitrate reduction to nitrite as a clinical characteristic for diagnosis for *B. subtilis* for several years (Gordon et al. 1973). As of the present time, it is known that *B. subtilis* possesses two different nitrate reductases: a respiratory one and an assimilatory one. Looking back, the nitrate reductase that was previously reported was from respiratory mechanism that is encoded by *narGHJI.* This is because it was later demonstrated that the enzyme is induced during anaerobiosis, rather than being controlled by the universal nitrogen system regulating the functions in *B. subtilis* and regulates the expression of genes that code for the assimilatory nitrate reductase (Nakano and Hulett 1997, Hoffmann et al. 1995).

Over the last three years, more thorough and methodical research has been done to investigate *B. subtilis* cells' capacity for anaerobic growth. From Priest's review, it appears that *B. subtilis* may expand and produce spores slowly in stringent anaerobic conditions even though it is classified as an aerobe. It develops anaerobically in the presence of carbohydrate (glucose) and nitrate using glycerol as the only carbon source as end point electron acceptor, as a consequence of which proliferation of *B. subtilis* in the absence of oxygen on nitrate is genuinely considered as respiratory (Nakano et al. 1997).

Pseudomonas aeruginosa exhibits to secrete chemicals which slows down the rate of oxygen transport. At least five terminal oxidases are encoded by this organism. The oxidase that is habitually produced (cytochrome cbb3-1) has an elevated oxygen affinity and is adapted in lesser levels of oxygen. The induction of cytochrome cbb3-2 occurs in extremely low oxygenation conditions, while low-affinity oxidases are mostly triggered by nutritional shortage or adverse stresses (Arai 2011, Arai et al. 2014). Anaerobic respiration of nitrate or nitrite (denitrification) is the supported process in *P. aeruginosa* in the event of a lack of oxygen. This process can also take place in hybrid metabolism under microaerobiosis. The Fnr-type transcription factor Anr, an apparent oxygen sensor at the summit of a complicated regulatory network, is primarily responsible for *P. aeruginosa*'s interaction with oxygen. While being regulated by Anr, another Fnr-like moderator called Dnr is shown to regulate the denitrification mechanism in addition to a limited part of other several genes activated in anaerobic condition. Thus, denitrification is governed by three regulatory regimes in *P. aeruginosa* namely, one Fnr/Crp-type regulator, Anr (PA1544), perceives the oxygen inadequacy, the two-component regulatory network, NarXL (PA3878; PA3879), regulates the level of nitrate, and another Fnr/Crp-type regulator, Dnr (PA0527), assess and regulates the production of NO (Acuna et al. 2017).

2.4.3 Hybrid Metabolism

In the context of anaerobiosis, a "hybrid mechanism" is often defined as a mixture of metabolic pathways or tactics that organisms use to adapt to low-oxygen situations. It illustrates how certain organisms are flexible and adaptive, using both anaerobic and aerobic processes to maximize energy generation and survival in a variety of oxygen environments. There are instances of hybrid metabolism, when the cytoplasm undergoes anaerobic activities, and the cell membrane undergoes aerobic respiration. The fumarate-nitrate reduction regulator (Fnr) and the ArcBA two-component system are the two regulators that govern the interaction between *E. coli* and oxygen (Rolfe et al. 2012, Unden et al. 1995). Fnr is a direct regulating system that controls the activation or repression of a wide range of operations. It detects oxygen by oxidizing a [4Fe-4S] cluster. ArcBA is a two-component system that senses the redox states of quinones that are membrane-bound and functions as an indirect oxygen regulator (Bettenbrock et al. 2014, Unden et al. 1995).

Microaerophilic organisms which thrive at oxygen partial pressures lower than one atmosphere may be at a competitive advantage over aerobic creatures. In order to maximize their energy metabolism and perform at their best in environments where oxygen availability is restricted but not completely absent, they frequently combine anaerobic and aerobic metabolic pathways. Numerous studies have been conducted on a few of the well-characterized microaerophiles, including species of the human pathogen Campylobacter. Additionally, deep subterranean sediments, marine sediments, coastal waters, and most recently, samples of soil, freshwater, and activated sludge have been shown to contain microaerophilic bacteria. However, little is known about the variety, range, and ecological significance of these species in the environment (Mannisto and Puhakka 2002).

2.5 Molecular Adaptation

2.5.1 Regulation of Gene Expression

In the absence of oxygen levels, cells undergo extensive changes in gene expression to optimize their metabolic pathways and maintain cellular homeostasis. Key transcription factors and regulatory proteins are activated to orchestrate the expression of genes involved in anaerobic metabolism, stress response, and energy conservation. Additionally, certain genes are repressed to minimize energy expenditure on processes that are dispensable in anaerobic conditions (Hwang et al. 2020, Lundin et al. 2015).

2.5.2 Characterization of Global Network of Gene Regulation During Aerobic-Anaerobic Transition in Anaerobes

The "ribonucleotide reductase (RNR)" network is the model system that we shall employ. RNRs are the prime enzymes required for production of deoxyribonucleotides (dNTPs) which are the main building block of DNA synthesis and repair. RNRs are thus of prime importance and suggested to be potential targets for antibacterial and cancer treatments because this mechanism is necessary for life. RNRs are classified into three categories (class I, that is further classified into Ia, Ib, and Ic; class II and class III) according to their structural strategy, metallocofactor, and radical generating process. They all share a radical-based mechanism. Class I RNRs are reliant on oxygen, class II RNRs are not dependent on oxygen, and class III RNRs are sensitive to oxygen. As a result of these distinctions, each RNR class behaves differently towards oxygen (Lundin et al. 2015). Bacterial microbes may encode an array of RNRs in multiple combinations in a way that corresponds to the environmental niches that each entity can occupy, but complex eukaryotes can only encode class Ia RNRs (Torrents 2014).

In *B. subtilis*, the homologous part of moaA, that is necessary for the production of the molybdenum cofactor for nitrate reductase, was shown to be the initial mutation that led to impaired nitrate respiration. Additionally, the study demonstrated that *B. subtilis* contains minimum of two nitrate reductases, each of which is essential for nitrate respiration. The primary identified genes responsible for nitrate respiration, namely *narGHJI, narK, fnr, resD*, and *resE*, were contributed by collaborative members of the "Bacillus genome sequencing project" (Glaser et al. 1995).

2.5.3 Gene Expression Changes

Molecular oxygen is mainly thought to be a key regulatory signal of metabolism in facultative anaerobic microbes. It may also function as a growth substrate, a toxic component, and a regulator for bacterial gene expression.

In *E. coli*, the regulation of respiration in the absence of oxygen via nitrate at the transcriptional level is due to the presence of two-component regulatory system known as NarX and NarL, which are nitrate sensor and respond to the regulatory system respectively. Similar substitute sensors and regulatory component (NarQ and NarP) which were recently found in *E. coli* recognized nitrite as a signal. Thus, the functional form of NarL produced in the occurrence of nitrate, known as phosphoryl-NarL (NarL-P), has the ability to function as a transcriptional promoter or repressor. Positive regulation is seen in genes related to nitrate respiration, including the structural and functional genes for nitrate reductase (narGHJl) and NADH-dependent nitrite reductase (nirB). Research has demonstrated that NarL is necessary for several enzymes associated with the nitrate suppression of respiration and fermentation in anaerobic condition, including fumarate reductase (frdABCD), DMSO-reductase (dmsABC), TMAO reductase (torA), fumarate/succinate antiport, and alcohol dehydrogenase (adhE) (Unden et al. 1994). The positive gene regulator FNR, which triggers transcription in the lack of 02, regulates the expression of anaerobic respiration and associated activities in *Escherichia coli*, yet shows no homology with the two component-regulatory systems (Unden and Trageser 1991, Spiro and Guest 1991).

Halobacterium species (strain NRC-1) grows best aerobically but also shows anaerobic growth which includes anaerobic respiration utilizing dimethyl sulphoxide (DMSO) or trimethylamine N-oxide (TMAO), substrate-level phosphorylation using arginine, and phototrophic development via the light-associated proton pump in their membrane. Anaerobic respiration necessitates the DMSO/TMAO reductase and a chaperone, which are expressed by the dmsREABCD operon, but anaerobic fermentation can be achieved by the arginine deiminase (ADI) cycle, which is represented by the arcRACB genes. The findings in this study indicated a rising transcriptional induction of gvp, bop, cba, dms, and arc genes under increasingly more restrictive settings implicated in induction of metabolic energy under hypoxic and anoxic conditions (Das Sarma et al. 2012).

Pseudomonas can grow anaerobically by denitrification or arginine degradation due to the presence of ANR gene, an analogue of FNR gene found in *E. coli* which acts as transcriptional activator of the anaerobically inducible pfl operon. The discovery that ANR is necessary for both heterologous expression of the *E. coli* pfl promoter system and anaerobic gene expression in Pseudomonas offers a great framework for analyzing the processes underlying ANF/FNR-dependent anaerobic regulation and also aids in clarifying the mechanisms governing the intricate pfl transcription system. Further, a pleiotropic loss of the ability to develop through anaerobic nitrate respiration or arginine breakdown was observed by mutations in the ANR gene (Sawers 1991).

2.5.4 Transcriptional Regulators Controlling the Synthesis of Enzymes Involved in Nitrate Respiration

Nitrate respiration in bacteria is adjourned by the following three parameters: Oxygen limitation, the presence of nitrate (NO_3^-) or derived compounds such as nitrite (NO_2) or Nitric Oxide (NO) in the medium, and an appropriate donor of electrons (i.e., organic carbon compounds). Combining these signals to modify the expression of genes involved in nitrate respiration is mostly dependent on a number of transcriptional regulators. Regulation of Dissimilatory Nitrate Reduction to Ammonium (DNRA) are regulated by FNR and two-component system, while the regulation of denitrification is directly co-regulated by the NarXL TCS and the DNR transcriptional regulators (Durand and Guillier 2021).

2.6 Physiological Adaptations of Extremophiles to Other Extreme Conditions

The term "extremophiles" refers to organisms that can survive under extreme conditions, including hydrothermal vents. Extremozymes are special enzymes utilized by these microorganisms, allowing them to survive in such harsh conditions. The two important extremophiles: thermophiles are organisms that thrive in high temperatures, while psychrophiles or cryophiles are species that thrive in cold temperatures (Kohli et al. 2020).

2.6.1 Psychrophiles and Stability

Microbial organisms that thrive in extreme environmental conditions and express optimum proliferation and reproduction at colder temperatures, and/or anaerobic condition often referred to as psychrophiles. Psychrophilic microbes that are anaerobic in nature are found in many different foods, soils, sewage systems, and other environments. On the other hand, it was not possible to identify and extract strictly anaerobic psychrophiles (Upadhyay and Stokes 1962). Psychrophiles include members of all three domains: bacteria, archaea, and eukaryotes. Primarily, the largest psychrophiles are polar fish that live under the ice (Eastman 2013). Because they can only complete their life cycle in glacial ice, glacier ice worms are also noteworthy (Roman Dial et al. 2012). Psychrophiles are the most prevalent extremophiles in terms of biomass, diversity, and dispersion (Feller 2013). These microbes are possible sources of cold-active enzymes which has gained interest in many commercial bioprocesses due to their increased stability and cold activity. Numerous cold-active enzymes, such as β-galactosidase, pectinase, proteases, amylases, xylanases, pullulanases, lipases, and β-mannanases, have gained uses throughout the food trade. Prospective enzyme expression in diverse hosts is made possible by recent advancements in the development of structural and functional proteins as well as enzyme technology, which ultimately result in enhanced catalytic and manufacturing capabilities (Kumari et al. 2021, Bakermans 2008).

Oxygen and heat have substantial impacts on the growth and metabolic processes of psychrophiles. It is possible that sugars that ferment at 20°C won't ferment at 0°C or won't create gases. Lower incubation temperatures prolong the lag, exponential, stationary, and terminal phases in both aerobic and anaerobic environments. The removal of oxygen shortens the maturation period, prolongs the cells' survival at low temperatures (but it speeds up their breakdown at 20°C and above), and, lastly, lessens the extent of growth. Aerobiologically, at 5°C, and anaerobically, at 25°C, maximum cell populations are reached. The availability or lack of oxygen has no discernible effect on any of these crucial temperatures with an optimal temperature of 20 and 35°C for growth (Upadhyay and Stokes 1962).

In cold environments, microorganisms have a defence against extreme temperatures interfering with protein stability and enzymatic function. The optimum temperature for growth of a Psychrophilic bacteria is 5°C, however it may stand temperature range 0–20°C by increasing the expression of enzymes that are sensitive to cold (Cavicchioli 2016, Salwan and Sharma 2020). Psychrophiles have complicated biochemical adaptations, including modified food transport pathways, intracellular ice formation, and cold denaturation of protein molecules, which enable them to sustain in adverse conditions. This improves membrane fluidity by increasing the amount of unsaturated lipids, expresses helicases, uses "cold shock proteins" to stabilize the nucleic acids (DNA/RNA), and makes use of pigments called carotenoids to shield the cell from the exposure to solar radiation (Rai and Kumar 2015, Bakermans et al. 2011). The enzymes adapt to the substantial decrease in the rate of chemical reaction by providing robust specificity even at low temperatures. This adaptation trait has been suggested to result from an extended selective process and is believed to be genetically encoded in the amino acid sequence. In psychrophilic bacteria, glycine betaine accumulating in the cytoplasm improved osmotolerance and cold tolerance (Feller 2013). The proteins of sub-zero-growing bacteria have more serine and less proline and acidic residues than those of mesophilic bacteria in order to promote protein fluidity at low temperatures (Raymond-Bouchard et al. 2018).

The abundance of specific essential proteins and metabolites in psychrophilic bacteria is crucial for their ability to adapt to cold settings. In frigid environments, they have evolved ways to sustain maximum translation and the folding of proteins. Many cold-adapted bacteria showed to have overexpressed ribosomal proteins and RNA helicases, the protein components responsible for maintaining RNA necessary for the translation process. Moreover, the active site of amino acids acquires conformational and dynamical strength by the reduction in the non-covalent bonds and interaction outside the catalytic loci, which maintains the enzymes' catalytic efficacy in low temperatures (Feller 2013, Jung et al. 2010). Furthermore, the adaptive characteristic of psychrophilic microorganisms and the conformational flexibility of the cold-active enzymes are substantially impacted by the chaperones, which include DnaK, GroEL, ribosome-bound Trigger Factor (TF), and RNA chaperones, such as cold-shock protein A. These chaperones have been stated to possess a key part in the folding of newly synthesised polypeptides (Kumari et al. 2021, Piette et al. 2011).

***Advantages and Application*:** "Cold-active or psychrophilic enzymes" are utilized in food manufacturing, and biotechnology. They are affordable, capable of catalyzing reactions without the need for extra heat assistance, and selectively in-activable with a fraction of heat intake. Their importance is anticipated to increase in the upcoming years as they prevent unwanted reactions and the loss of volatile components. Fruit and alcoholic beverages, chocolate products, sweeteners, dairy products, and baked goods, are a few examples (Al-Ghanayem and Joseph 2020, Kuddus 2018).

2.6.2 Thermophiles and Stability

Thermophiles are microorganisms that have an optimum temperature for growth (OGT) of 60 to 80°C; on the other hand, hyperthermophiles are microbes that grow optimally at temperatures higher than 80°C. The majority of these organisms are bacteria and archaea. Anaerobic thermophilic Archaea are undoubtedly the most "extreme" in terms of habitable environments among anaerobic and thermophilic bacteria. In the universal phylogenetic tree, they stand for the least evolved and interior-most branches. They may have been the first living things on this planet since they frequently employ substrates, which are assumed to have been predominant in the primordial terrestrial makeup. Many features of protein, membrane, and nucleic acid structure have been clarified by studies into how they control thermostability at the structural level. Nevertheless, the principles of thermostability are still not fully understood. Future practical uses will depend on the improvement of genetic techniques for using these organisms (Canganella and Wiegel 2014). Anaerobic thermophiles have developed a number of adaptations that enable them to endure temperatures that other species cannot tolerate because of the strain of surviving at such high temperatures. These characteristics comprise greater turnover rates for different protein enzymes, thermally stable membrane proteins, and a distinct lipid makeup of the membrane. The plasma membrane that surrounds the body is one of the most crucial components in maintaining homeostasis within the organism. Anaerobic thermophiles need to stabilize not just the plasma membrane but also their proteins, DNA, RNA, and ATP at high temperatures (Canganella and Wiegel 2014).

2.6.3 Morphological and Structural Changes

In addition to metabolic and genetic adaptations, some organisms undergo morphological and structural changes in response to anaerobiosis. For instance, certain microorganisms alter their cell membrane composition or develop specialized structures to enhance gas exchange and nutrient uptake in oxygen-deprived environments. These morphological adaptations play a critical role in facilitating cellular survival under anaerobic conditions. Understanding the intricate interplay between cellular morphology and anaerobic adaptation provides valuable knowledge for diverse fields, including environmental studies, bioremediation, and biotechnology. These insights can inspire innovative approaches for engineering microorganisms with enhanced anaerobic tolerance and metabolic capabilities, as well as for designing solutions to address environmental challenges related to oxygen-deprived ecosystems (Lowe et al. 1993, Gupta et al. 2017).

2.6.4 Changes in Cell Morphology/Structure

In the face of oxygen limitation, cells undergo remarkable changes in morphology to adapt to anaerobiosis. One of the striking morphological adaptations observed in various organisms is the alteration of cell membrane composition. This change in membrane structure and composition enables cells to maintain integrity and function in the absence of oxygen, facilitating the exchange of gases and nutrients essential for survival. Furthermore, certain microorganisms develop specialized structures, such as gas vesicles or biofilms, to enhance gas exchange and create microenvironments that support their growth and metabolism under anaerobic conditions. These structural modifications contribute to the ability of organisms to thrive in oxygen-deprived environments and represent a fascinating aspect of cellular adaptation to anaerobiosis (Rowlett et al. 2017, Son et al. 2023).

In addition to changes in cell morphology, anaerobic conditions can also lead to alterations in cell structure at the molecular level. One noteworthy alteration is the modification of cell wall composition and structure to withstand the challenges posed by oxygen limitation. Certain organisms, particularly bacteria and fungi, reinforce their cell walls to maintain stability and protect against environmental stressors encountered in anaerobic environments. Furthermore, the cytoskeletal organization within cells may undergo modifications to support cellular functions under oxygen-deprived conditions. These changes in cytoskeletal dynamics can influence the spatial arrangement of organelles and cellular components, contributing to the overall resilience of cells facing anaerobiosis. The adaptations in cell structure represent an essential component of the multifaceted responses exhibited by organisms in anaerobic environments (Rowlett et al. 2017, Sarkar et al. 2009).

2.6.5 Formation of Specialized Structures in Anaerobic Environments

In anaerobic environments, many organisms undergo the formation of specialized structures that enable them to thrive and adapt to the absence of oxygen. One remarkable example of such structures is the development of gas vesicles in certain microorganisms. Gas vesicles are proteinaceous, gas-filled structures that confer buoyancy to cells, allowing them to position themselves optimally within their habitat to access essential nutrients and light while avoiding toxic by-products. Additionally, the formation of biofilms represents another fascinating adaptation to anaerobiosis. Biofilms are complex communities of microorganisms encased in a self-produced extracellular matrix, providing protection, and promoting cooperative interactions between cells. In anaerobic conditions, biofilms offer a survival advantage by facilitating the exchange of metabolites and creating microenvironments that support diverse metabolic activities, contributing to the resilience of organisms in oxygen-deprived settings (Feng et al. 2024).

Insights into the mechanisms underlying the development of these structures can inspire innovative approaches in bioremediation, wastewater treatment, and bioenergy production. Furthermore, exploiting the unique properties of these structures may lead to the design of advanced materials and technologies with applications in biotechnology and environmental engineering (Uruén et al. 2020, Pfeifer 2022).

2.6.6 Impact on Cellular Organelles

Anaerobic conditions exert significant impacts on cellular organelles, leading to diverse adaptations that enable cells to maintain essential functions in the absence of oxygen. One of the prominent effects is observed in the mitochondria, the powerhouse of the cell responsible for energy production. In anaerobic environments, the function of mitochondria is altered, leading to the activation of alternative pathways for ATP generation and metabolic adjustments to cope with the absence of oxygen (Hackstein et al. 2001). Additionally, the endoplasmic reticulum, involved in protein synthesis and lipid metabolism, undergoes modifications to support the production of proteins essential for anaerobic adaptation. Cells may also exhibit changes in the structure and function of the Golgi apparatus, which plays a crucial role in processing and transporting cellular macromolecules. These adaptations collectively contribute to the maintenance of essential cellular processes under oxygen-deprived conditions (Hackstein et al. 2001, Jalil et al. 2023, Stairs et al. 2015).

2.6.7 Mitochondrial Adaptations to Anaerobic Environments

Mitochondria, crucial for energy production in aerobic conditions, undergo significant adaptations to ensure cellular energy supply in anaerobic environments. One notable adaptation is the activation of alternative pathways for ATP generation, such as substrate-level phosphorylation, which allows cells to generate ATP without relying on oxygen-dependent processes like oxidative phosphorylation (Stairs et al. 2015). Additionally, the mitochondria may undergo structural and functional adjustments to optimize their performance under low oxygen conditions, ensuring that essential energy needs are met. Moreover, in anaerobic settings, mitochondria exhibit changes in their metabolic activities, such as the utilization of alternate substrates for energy production to compensate for the limited availability of oxygen. These metabolic adaptations enable cells to sustain vital biological processes even in the absence of oxygen, reflecting the remarkable flexibility and resilience of cellular systems in response to environmental changes (Hackstein et al. 2001, Stairs et al. 2015).

2.6.8 Changes in Endoplasmic Reticulum Function in Anaerobic Environments

The endoplasmic reticulum undergoes notable changes in function to support cellular adaptation to anaerobic environments. As oxygen becomes limited, the ER adjusts its activities to ensure the production of proteins essential for the anaerobic response. This may involve the upregulation of specific chaperone proteins and the modulation of protein folding processes to meet the demands of anaerobic adaptation (Jalil et al. 2023). Furthermore, the ER exhibits alterations in lipid metabolism to accommodate the production of lipids crucial for maintaining cell membrane integrity and function under oxygen-deprived conditions. These adjustments enable cells to sustain essential cellular processes and ensure the integrity of intracellular membranes (Jalil et al. 2023).

2.6.9 Solvent Tolerance Mechanisms in Anaerobic Environments

In anaerobic environments, microorganisms have developed various mechanisms to tolerate and thrive in the presence of solvents. Solvents can disrupt cellular membranes and impair essential cellular processes, posing a significant challenge to microbial survival. To overcome this, anaerobic microorganisms have evolved unique strategies to mitigate the detrimental effects of solvents and maintain their cellular integrity. One of the key mechanisms is the modification of membrane composition, where cells alter the lipid content and structure of their membranes to enhance resistance to solvents. This adaptive response reduces the permeability of the cell membrane to solvents, thereby shielding the intracellular components from damage and preserving cellular function (Schalck et al. 2021). Moreover, anaerobic microorganisms may employ efflux pumps to actively extrude solvents from the cell, preventing their accumulation to toxic levels. These efflux systems actively transport solvents out of the cell, contributing to the overall tolerance of anaerobic microorganisms to solvent exposure. Understanding the solvent tolerance mechanisms in anaerobic environments is not only crucial for elucidating fundamental microbial adaptations but also holds practical significance in industrial biotechnology and environmental remediation. By leveraging the knowledge of these mechanisms, researchers can develop strategies to enhance solvent tolerance in microbial strains used for bioprocessing and bioremediation, as well as design novel approaches for the detoxification of solvent-contaminated environments (Koenig et al. 2014).

2.6.10 Role of Cryoprotectants and Antifreeze Proteins in Anaerobic Environments

In addition to solvent tolerance mechanisms, anaerobic microorganisms also utilize cryoprotectants and antifreeze proteins to thrive in their environments. These protective molecules play a crucial role in enhancing cellular survival and function under low-temperature conditions, which are often prevalent in anaerobic settings. Cryoprotectants, such as sugars and polyols, function by reducing the formation of ice crystals within cells, thereby preventing cellular damage caused by freezing. These molecules effectively lower the freezing point of cellular fluids, enabling microorganisms to maintain metabolic activities and essential cellular processes even in sub-zero temperatures. On the other hand, antifreeze proteins actively bind to ice crystals and inhibit their growth, further safeguarding cellular structures from the detrimental effects of low temperatures. By preventing ice formation and growth, antifreeze proteins contribute to the overall resilience of anaerobic microorganisms in cold environments (Ekpo et al. 2022).

Understanding the role of cryoprotectants and antifreeze proteins in anaerobic environments provides valuable insights into the strategies employed by microorganisms to thrive in extreme conditions. Moreover, this knowledge holds potential for applications in various industries, including cryopreservation of biological materials, cold-tolerant biotechnological processes, and the development of novel biopharmaceuticals with enhanced stability under low-temperature conditions (Białkowska et al. 2020).

2.7 Future Perspectives

Anaerobiosis, the absence of oxygen, presents a significant challenge for many organisms. In summary, the adaptations of cellular organelles and microorganisms to anaerobic environments reveal the remarkable resilience and flexibility of biological systems in response to environmental challenges. By delving into these mechanisms, researchers can not only deepen our understanding of fundamental biological processes but also pave the way for innovative applications in healthcare, biotechnology, and environmental sustainability.

The horizon is full of prospects for the further uses of anaerobic microorganisms, given the significant roles that they have been documented to serve. The main hurdles to the broader use of anaerobic bacteria are their unculturable nature, their adverse conditions, which are still unexplored, and our incomplete understanding of their metabolic and genetic potential. In the future, novel anaerobic bacteria from unidentified severe habitats may be identified using RAST genomic annotation pipeline-based tools, ANNs (Artificial Neural Networks), as well as metagenomic-assisted approaches. Anaerobic bacteria can be made more efficient for a variety of purposes through genetic engineering. Understanding the intricate mechanisms underlying the adaptation to anaerobiosis is crucial for various fields, including biotechnology, medical research, and environmental science. By unravelling the physiological and molecular responses to anaerobiosis, researchers can gain insights into optimizing biofuel production, developing strategies to combat anaerobic infections, and deciphering the impact of anaerobiosis on ecosystems.

References

Acuña, B., Manuel, J., Timmis, K. N., Jahn, M. and Jahn, D. (2017). Protein complex formation during denitrification by *Pseudomonas aeruginosa*. Microbial Biotechnology 10(6): 1523–34.

Alexeeva, S., Hellingwerf, K. J. and Mattos, M. J. T. (2002). Quantitative assessment of oxygen availability: perceived aerobiosis and its effect on flux distribution in the respiratory chain of *Escherichia coli*. Journal of Bacteriology 184(5): 1402–6.

Al-Ghanayem, A. A. and Joseph, B. (2020). Current prospective in using cold-active enzymes as eco-friendly detergent additive. Applied Microbiology and Biotechnology 104(7): 2871–82.

Allen, J. F. and Martin, W. (2007). Evolutionary biology: out of thin air. Nature 445(7128): 610–12.

Arai, H. (2011). Regulation and function of versatile aerobic and anaerobic respiratory metabolism in *Pseudomonas aeruginosa*. Frontiers in Microbiology 2.

Arai, H., Kawakami, T., Osamura, T., Hirai, T., Sakai, Y. and Ishii, M. (2014). Enzymatic characterization and *in vivo* function of five terminal oxidases in *Pseudomonas aeruginosa*. Journal of Bacteriology 196(24): 4206–15.

Bakermans, C. (2008). Limits for microbial life at subzero temperatures. pp. 17–28. *In*: Psychrophiles: From Biodiversity to Biotechnology, edited by Rosa Margesin, Franz Schinner, Jean-Claude Marx, and Charles Gerday, Berlin, Heidelberg: Springer.

Bakermans, C. P., Bergholz, W., Rodrigues, D. F., Vishnivetskaya, T. A., Ayala-del-Río, H. L. and Tiedje, J. M. (2011). Genomic and expression analyses of cold-adapted microorganisms. pp. 126–55. *In*: Polar Microbiology.

Bettenbrock, K., Bai, H., Ederer, M., Green, J., Hellingwerf, K. J., Holcombe, M. et al. (2014). Towards a systems level understanding of the oxygen response of *Escherichia coli*. Advances in Microbial Physiology 64: 65–114.

Białkowska, A., Majewska, E., Olczak, A. and Clapa, A. T. (2020). Ice binding proteins: diverse biological roles and applications in different types of industry. Biomolecules 10(2): 274.

Bragg, P. D. (1980). The respiratory system of *Escherichia coli*. *In*: Diversity of Bacterial Respiratory Systems. CRC Press.

Canganella, F. and Wiegel, J. (2014). Anaerobic thermophiles. Life (Basel, Switzerland) 4(1): 77–104.

Cavicchioli, R. (2016). On the concept of a psychrophile. The ISME Journal 10(4): 793–95.

Clark, D. P. (1989). The fermentation pathways of *Escherichia coli*. FEMS Microbiology Reviews 5(3): 223–34.

Cole, L. A. (2016). Chapter 13 - Evolution of chemical, prokaryotic, and eukaryotic life. pp. 93–99. *In*: Biology of Life, edited by Laurence A. Cole. Academic Press.

Cooper, G. M. (2000). The Origin and Evolution of Cells. In The Cell: A Molecular Approach. 2nd Edition. Sinauer Associates. https://www.ncbi.nlm.nih.gov/books/NBK9841/.

Das, S. P., Zamora, R. C., Müller, J. A. and Shsrma, S. D. (2012). Genome-wide responses of the model archaeon Halobacterium sp. strain NRC-1 to oxygen limitation. Journal of Bacteriology 194(20): 5530–37.

Donnelly, M. I., Millard, C. S., Clark, D. P., Chen, M. J. and Rathke, J. W. (1998). A novel fermentation pathway in an *Escherichia coli* mutant producing succinic acid, acetic acid, and ethanol. Applied Biochemistry and Biotechnology 70(1): 187–98.

Durand, S. and Guillier, M. (2021). Transcriptional and post-transcriptional control of the nitrate respiration in bacteria. Frontiers in Molecular Biosciences 8: 667758.

Eastman, J. T. (2013). Antarctic Fish Biology: Evolution in a Unique Environment. Academic Press.

Ekpo, M. D., Xie, J., Hu, Y., Liu, X., Liu, F., Xiang, J. et al. (2022). Antifreeze proteins: novel applications and navigation towards their clinical application in cryobanking. International Journal of Molecular Sciences 23(5): 2639.

Feller, G. (2013). Psychrophilic enzymes: from folding to function and biotechnology. Scientifica. 512840.

Feng, R. J. L., Goh, M. C., Du, M. and Chen, Z. (2024). Advances in the application of gas vesicles in medical imaging and disease treatment. Journal of Biological Engineering 18(1): 41.

Ghasemi, A. (2022). Quantitative aspects of nitric oxide production from nitrate and nitrite. EXCLI Journal 21(February): 470–86.

Glaser, P. A. D., Kunst, F., Zuber, P. and Nakano, M. M. (1995). Identification and isolation of a gene required for nitrate assimilation and anaerobic growth of *Bacillus subtilis*. Journal of Bacteriology 177(4): 1112–15.

Gordon, R. E., Haynes, W. C. and Hor-Nay Pang, C. (1973). The Genus *Bacillus*. Agricultural Research Service, U.S. Department of Agriculture.

Gupta, A., Gupta, R. and Singh, R. L. (2017). Microbes and Environment. Principles and Applications of Environmental Biotechnology for a Sustainable Future. Nature Publishing Group. 43.

Hackstein, J. H., Akhmanova, A., Voncken, F., van Hoek, A., van Alen, T., Boxma, B. et al. (2001). Hydrogenosomes: convergent adaptations of mitochondria to anaerobic environments. Zoology (Jena, Germany) 104(3–4): 290–302.

Hentges, D. J. (1996). Anaerobes: General Characteristics. In Medical Microbiology, edited by Samuel Baron, 4th ed. Galveston (TX): University of Texas Medical Branch at Galveston.

Hoffmann, T., Troup, B., Szabo, A., Hungerer, C. and Jahn, D. (1995). The anaerobic life of *Bacillus subtilis*: cloning of the genes encoding the respiratory nitrate reductase system. FEMS Microbiology Letters 131(2): 219–25.

Hwang, J. H., Yu, S., Lee, B. and Lee, D. H. (2020). Modulation of energy metabolism is important for low-oxygen stress adaptation in Brassicaceae species. International Journal of Molecular Sciences 21(5): 1787.

Jalil, A. T., Abdulhadi, M. A., Alkubaisy, S. A., Thejeel, S. H., Essa, I. M., Merza, M. S. et al. (2023). The role of endoplasmic reticulum stress in promoting aerobic glycolysis in cancer cells: an overview. Pathology, Research and Practice 251: 154905.

Jones, S. A., Chowdhury, F. Z., Fabich, A. J., Anderson, A., Schreiner, D. M., House, A. L. et al. (2007). Respiration of *Escherichia coli* in the mouse intestine. Infection and Immunity 75(10): 4891–99.

Jung, Y. H., Yi, J. Y., Jung, H. J., Lee, Y. K., Lee, H. K., Naicker, M. C. et al. (2010). Overexpression of cold shock protein A of Psychromonas Arctica KOPRI 22215 confers cold-resistance. The Protein Journal 29(2): 136–42.

Jurtshuk, P. (1996). Bacterial Metabolism. Medical Microbiology, edited by Samuel Baron, 4th ed. Galveston (TX): University of Texas Medical Branch at Galveston. http://www.ncbi.nlm.nih.gov/books/NBK7919/.

Koenig, J. C., Groissmeier, K. D. and Manefield, M. J. (2014). Tolerance of anaerobic bacteria to chlorinated solvents. Microbes and Environments 29(1): 23–30.

Kohli, I., Joshi, N. C., Mohapatra, S. and Varma, A. (2020). Extremophile—an adaptive strategy for extreme conditions and applications. Current Genomics 21(2): 96–110.

Kuddus, M. (2018). Cold-active enzymes in food biotechnology: an updated mini review. Journal of Applied Biology and Biotechnology 6(3): 58–63.

Kumari, M., Padhi, S., Sharma, S., Phukon, L. C., Singh, S. P. and Rai, A. K. (2021). Biotechnological potential of psychrophilic microorganisms as the source of cold-active enzymes in food processing applications. 3 Biotech 11(11): 479.

Lowe, S. E., Jain, M. K. and Zeikus, J. G. (1993). Biology, ecology, and biotechnological applications of anaerobic bacteria adapted to environmental stresses in temperature, pH, salinity, or substrates. Microbiological Reviews 57(2): 451–509.

Lundin, D., Berggren, G., Logan, D. T. and Sjöberg, B. M. (2015). The origin and evolution of ribonucleotide reduction. Life 5(1). Multidisciplinary Digital Publishing Institute: 604–36.

Mannisto, M. K. and Puhakka, J. A. (2002). Psychrotolerant and microaerophilic bacteria in boreal groundwater. FEMS Microbiology Ecology 41(1): 9–16.

Nakano, M. M., Dailly, Y. P., Zuber, P. and Clark, D. P. (1997). Characterization of anaerobic fermentative growth of *Bacillus subtilis*: Identification of fermentation end products and genes required for growth. Journal of Bacteriology 179(21): 6749–55.

Nakano, M. M. and Marion Hulett, F. (1997). Adaptation of *Bacillus subtilis* to oxygen limitation. FEMS Microbiology Letters 157(1): 1–7.

Pedraz, L., Cabra, N. B. and Torrents, E. (2020). Gradual adaptation of facultative anaerobic pathogens to microaerobic and anaerobic conditions. The FASEB Journal 34(2): 2912–28.

Pfeifer, F. (2022). Recent advances in the study of gas vesicle proteins and application of gas vesicles in biomedical research. Life 12(9). Multidisciplinary Digital Publishing Institute: 1455.

Pieper, L. M., Spanogiannopoulos, P., Volk, R. F., Miller, C. J., Wright, A. T. and Turnbaugh, P. J. (2023). Global anaerobic metabolism regulator Fnr is necessary for the degradation of food dyes and drugs by *Escherichia coli*. mBio. 14(5): e01573–23.

Pierre, B. J., Bohin, A. and Schaeffer, P. (1976). Increased nitrate reductase A activity as a sign of membrane alteration in early blocked asporogenous mutants of *Bacillus subtilis*. Biochimie 58(1): 99–108.

Piette, F., Struvay, C. and Feller, G. (2011). The protein folding challenge in psychrophiles: facts and current issues. Environmental Microbiology 13(8): 1924–33.

Rai, A. and Kumar, R. (2015). Potential of microbial bio-resources of sikkim himalayan region. ENVIS Bulletin Himalayan Ecology 23: 99–105.

Raymond, B., Goordial, I. J., Zolotarov, Y., Ronholm, J., Stromvik, M., C. et al. (2018). Conserved genomic and amino acid traits of cold adaptation in subzero-growing arctic permafrost bacteria. FEMS Microbiology Ecology 94(4): fiy023.

Rolfe, M. D., Ocone, A., Stapleton, M. R., Hall, S., Trotter, E. W., Poole, R. K. et al. (2012). Systems analysis of transcription factor activities in environments with stable and dynamic oxygen concentrations. Open Biology 2(7): 120091.

Roman, D. C., Dial, R. J., Saunders, R., Lang, S. A., Lee, B., Wimberger, P. et al. (2012). Historical biogeography of the North American Glacier Ice Worm, Mesenchytraeus Solifugus (Annelida: Oligochaeta: Enchytraeidae). Molecular Phylogenetics and Evolution 63(3): 577–84.

Rowlett, V. W., Mallampalli, V. K. P. S., Karlstaedt, A., Dowhan, W., Taegtmeyer, H., Margolin, W. et al. (2017). Impact of membrane phospholipid alterations in *Escherichia coli* on cellular function and bacterial stress adaptation. Journal of Bacteriology 199(13): e00849–16.

Sabra, W., Kim, E. J. and Zeng, A. P. (2002). Physiological responses of *Pseudomonas aeruginosa* PAO1 to oxidative stress in controlled microaerobic and aerobic cultures. Microbiology (Reading, England) 148(Pt 10): 3195–3202.

Salwan, R. and Sharma, V. (2020). Physiological and Biotechnological Aspects of Extremophiles. Academic Press.

Sarkar, P., Bosneaga, E. and Auer, M. (2009). Plant cell walls throughout evolution: towards a molecular understanding of their design principles. Journal of Experimental Botany 60(13): 3615–35.

Sawers, R. G. (1991). Identification and molecular characterization of a transcriptional regulator from *Pseudomonas aeruginosa* PAO1 exhibiting structural and functional similarity to the FNR protein of *Escherichia coli*. Molecular Microbiology 5(6): 1469–81.

Schalck, T., den Bergh, B. V. and Michiels, J. (2021). Increasing solvent tolerance to improve microbial production of alcohols, terpenoids and aromatics. Microorganisms 9(2): 249.

Schirawski, J. and Unden, G. (1995). Anaerobic respiration of Bacillus Macerans with Fumarate, TMAO, nitrate and nitrite and regulation of the pathways by oxygen and nitrate. Archives of Microbiology 163(2): 148–54.

Son, Y., Min, J., Shin, Y. and Park, W. (2023). Morphological and physiological adaptations of Psychrophilic Pseudarthrobacter Psychrotolerans YJ56 under Temperature Stress. Scientific Reports 13(1). Nature Publishing Group: 14970.

Spiro, S. and Guest, J. R. (1991). Adaptive responses to oxygen limitation in *Escherichia coli*. Trends in Biochemical Sciences 16(8): 310–14.

Stairs, C. W., Leger, M. M. and Roger, A. J. (2015). Diversity and origins of anaerobic metabolism in mitochondria and related organelles. Philosophical Transactions of the Royal Society B: Biological Sciences 370(1678): 20140326.

Tenaillon, O., Skurnik, D., Picard, B. and Denamur, E. (2010). The population genetics of commensal *Escherichia coli*. Nature Reviews. Microbiology 8(3): 207–17.

Todgham, A. E. and Stillman, J. H. (2013). Physiological responses to shifts in multiple environmental stressors: relevance in a changing world. Integrative and Comparative Biology 53(4): 539–44.

Torrents, E. (2014). Ribonucleotide reductases: essential enzymes for bacterial life. Frontiers in Cellular and Infection Microbiology 4: 52.

Unden, G. and Trageser, M. (1991). Oxygen regulated gene expression in *Escherichia coli*: Control of anaerobic respiration by the FNR protein. Antonie Van Leeuwenhoek 59(2): 65–76.

Unden, G., Becker, S., Bongaerts, J., Schirawski, J. and Six, S. (1994). Oxygen regulated gene expression in facultatively anaerobic bacteria. Antonie Van Leeuwenhoek 66(1–3): 3–22.

Unden, G., Bongaerts, S. B. J., Holighaus, G., Schirawski, J. and Six, S. (1995). O_2-Sensing and O_2-dependent gene regulation in facultatively anaerobic bacteria. Archives of Microbiology 164(2): 81–90.

Upadhyay, J. and Stokes, J. L. (1962). Anaerobic growth of psychrophilic bacteria. Journal of Bacteriology 83(2): 270–75.

Uruén, C., Escuin, G. C., Tommassen, J., Mainar-Jaime, R. C. and Arenas, J. (2020). Biofilms as promoters of bacterial antibiotic resistance and tolerance. Antibiotics 10(1): 3.

Worlitzsch, D., Tarran, R., Ulrich, M., Schwab, U., Cekici, A., Meyer, K. C. et al. (2002). Effects of reduced mucus oxygen concentration in airway *Pseudomonas* infections of cystic fibrosis patients. The Journal of Clinical Investigation 109(3): 317–25.

Zhu, D., Adebisi, W. A., Ahmad, F., Sethupathy, S., Danso, B. and Sun, J. (2020). Recent development of extremophilic bacteria and their application in biorefinery. Frontiers in Bioengineering and Biotechnology 8.

Chapter 3

Taxonomy and Evolution of Anaerobic Bacteria and Methanogenic Archaea

Sanjana, N.S., Sridevi, K.S., Akshaya, J., Ananya, K. and *Bhagwan Narayan Rekadwad**

3.1 Introduction

In 1860, Louis Pasteur's discovery of anaerobes opened a door of the era anaerobic microbiology. In humans, these bacterial communities are prevalent within the gastrointestinal tract (Rupnik et al. 2021). Methanogenic Archaea, on the other hand, are essential for the fermentative decomposition of organic matter (Minh). There are five primary groups of anaerobic bacteria: *Actinomyces, Bifidobacterium, Fusobacterium, Propionibacterium, Clostridium, Bacteroides,* and *Prevotella* (Jousimies-Somer 1997). Anaerobic archaea, also known as methanogens, proliferate by releasing methane gas (Vítězová et al. 2020). Decades of research into microbial physiology have been driven by these microorganisms and their unique metabolic processes, expanding our understanding of how microbes store energy for growth (Barker 1936). Methanogens are crucial to human well-being, playing significant physiological roles. Moreover, other anaerobic organisms are vital to the global carbon cycle. Archaea contribute significantly to this cycle by catalyzing the production of methane during the final stage of anaerobic biomass breakdown (Hoegenauer et al. 2022).

Though they are found in a variety of biological niches, anaerobic bacteria are essential to the maintenance of the global cycles of carbon, nitrogen, and

MicrobeAI Lab, Division of Microbiology and Biotechnology, Yenepoya Research Centre, Yenepoya (Deemed to be University), University Road, Deralakatte, Mangalore – 575018, India.

* Corresponding author: rekadwad@gmail.com

DOI: 10.1201/9781003506898-3

sulfur as well as the breakdown of persistent chemicals. As a result, they are now valuable components in environmental and industrial biotechnology (Hatti-Kaul and Mattiasson 2016). A frequent use of anaerobic biotechnology is the treatment of highly concentrated wastewaters from various industries. The anaerobic environment in which many pollutants are converted into energy-rich biogas is the basis for this biotechnology, which relies on the interactions and cooperation of microorganisms (Vítězová et al. 2020). The characteristics of wastewater differ among businesses and have a substantial impact on the microbiome makeup within the anaerobic reactor. Methanogenic archaea are essential to the anaerobic digestion of wastewater. *Methanosarcina* sp. and *Methanotrix* sp. are the most prevalent acetoclastic methanogens in anaerobic reactors used to treat industrial effluent (Abken et al. 1998). Anaerobic wastewater treatment is an extremely effective and environmentally friendly alternative to aerobic technology. Numerous sources of anaerobic wastewater treatment exist, making it a valuable tool for sustainable waste management.

Methanogens are anaerobic archaea that grow by producing methane gas. These microbes and their unique metabolism have inspired decades of microbial physiology research, continuously expanding our understanding of how microbes conserve energy for growth. Anaerobic archaea, also known as methanogens, proliferate by releasing methane gas. Decades of microbial physiology research has been driven by these microorganisms and their unique metabolism, which continues to push the boundaries of our knowledge about how microbes store energy for growth (Hoegenauer et al. 2022). Previous genomic investigations had shown that anaerobic organisms are capable of producing secondary metabolites and had identified clostridia from soil as a viable source for the discovery of new natural products (Letzel et al. 2013). Anaerobes have previously yielded novel chemicals with distinct molecular scaffolds or functions; additional instances are anticipated as a result of using genome mining to connect Biosynthesis Gene Clusters (BGCs) to their metabolites. Apart from the potential chemical space that anaerobes offer, the production of these distinct compounds likely involves intriguing enzymes and pathways, some of which remain unidentified (Li et al. 2019). These organisms are well known to produce toxins, thrive in anaerobic environments, and work in concert with anti-cancer medications to cause tumor shrinkage and inhibit metastasis. It may be possible to exploit these characteristics in addition to traditional cancer treatments. To the best of our knowledge, no publications have been made about the most prevalent bacteria that target tumors, with particular attention paid to facultative anaerobes like *Salmonella* sp., *Listeria monocytogenes, Lactobacillus* sp., *Escherichia coli*, *Corynebacterium diphtheriae*, and *Pseudomonas* sp., as well as obligate anaerobes like *Clostridium* sp. and *Bifidobacterium* sp. (Dróżdż et al. 2020).

3.2 Taxonomy Anaerobes and the Importance of 16S rRNA Gene Sequencing

Taxonomy, a fundamental branch of biology, is concerned with the classification and systematic organization of organisms. It encompasses both evolutionary and phylogenetic relationships (Ferraz Helene et al. 2022). Among the diverse groups of microorganisms, anaerobic bacteria thrive in environments devoid of oxygen (Moore et al. 2010). The 16S rRNA gene has emerged as a valuable genetic marker for studying bacterial phylogeny and taxonomy. Its sequence analysis enables the identification of novel genera and species that were previously uncharacterized (Janda and Abbott 2007).

In recent years, phylogenetic classification of prokaryotes has heavily relied on the 16S rRNA gene sequence. Although this gene can be horizontally transferred between different species, it retains sufficient specificity to distinguish individual species (Kitahara and Miyazaki 2013). By comparing the complete 16S rRNA gene sequences of various prokaryotic strains, researchers can establish taxonomic relationships based on sequence similarity. A commonly used threshold for this comparison is 98.65% (Kim and Chun 2014). The advent of Polymerase Chain Reaction (PCR) and Sanger sequencing techniques significantly simplified the determination of the 16S rRNA gene's primary structure. Classification and identification of bacteria primarily hinge on comparing these sequences to existing databases. The 16S rRNA gene sequence serves two primary purposes: calculating pairwise sequence similarities and conducting phylogenetic analyses of multiple sequence alignments. Phylogenetic analysis elucidates the evolutionary relationships between bacteria and related taxa (Kim and Chun 2014). Moreover, automated systems like MALDI-TOF and Vitek 2 have greatly simplified and fastened the identification of anaerobic bacteria (Ayesha et al. 2022). These microorganisms can be either Gram-positive or Gram-negative, encompassing a wide range of species. For instance, the strictly anaerobic or aerotolerant spore-forming bacilli found in soil and the gastrointestinal tract of humans and animals, including Clostridia species, represent a significant group of obligate anaerobes (Wells and Wilkins 1996).

3.2.1 The Dangers of Clostridium Species

Certain Clostridium species, such as *C. difficile*, *C. septicum*, and *C. perfringens*, are notorious for causing gas gangrene, a severe and potentially fatal condition (Noor and Khetarpal 2023).

3.2.2 Actinomyces: A Group of Non-spore Forming Anaerobes

Actinomyces, another important group of anaerobic bacteria, are non-spore forming, non-acid-fast bacilli. They are gram-positive bacteria that are commonly found in the human body, particularly in the mouth, urogenital tract, and other areas. While some Actinomyces species can cause infections, they are generally considered part of the normal human microbiota (Sharma et al. 2024).

3.2.3 Facultative Anaerobes: Adapting to Different Environments

Facultative anaerobes are microorganisms capable of thriving in both oxygen-rich and oxygen-depleted environments. These versatile organisms possess a variety of oxygen-independent electron transport chains, enabling them to switch between aerobic and anaerobic respiration (Simon 2013). A prominent example of facultative anaerobes is the Enterobacteriaceae family, a group of Gram-negative bacilli that commonly inhabit the gastrointestinal tracts of animals and humans. These bacteria are characterized by their outer membrane, which acts as a barrier against certain antibiotics, including vancomycin and daptomycin. Notable members of the Enterobacteriaceae family include *Escherichia coli*, *Klebsiella pneumoniae*, and *Enterobacter cloacae* (Jones and Howe 2014). In contrast, strict anaerobes require oxygen-free conditions for growth. Bacteria like *Pasteurella* sp., *Prevotella* sp., and *Porphyromonas* sp. fall into this category. These anaerobes can breach mucosal barriers and invade subcutaneous tissues, leading to infections (Sykes 2014).

3.3 Taxonomy of Methanogenic Archaea

Archaea, a domain of prokaryotic organisms, were predominantly found in extreme environments such as hot springs and hydrothermal vents, earning them the nickname "extremophiles." While they could also be found in more common habitats like soil and the human gut, their unique adaptations to harsh conditions set them apart. These adaptations, including stable proteins and enzymes under extreme temperatures and salinities, played a crucial role in global carbon and nitrogen cycles (Aparici-Carratalá et al. 2023). Archaea exhibited a diverse range of shapes and sizes, varying from 0.7 μm to 4 μm. Their morphology often resembled bacteria, with cocci and rod-shaped forms being common (van Wolferen et al. 2022). Key characteristics that distinguished archaea from bacteria included the absence of peptidoglycan in their cell walls, the presence of ether-linked lipids, their unique habitats, and distinct tRNA and rRNA sequences (Woese et al. 1978).

3.3.1 Methanogens

Methanogens were a prominent subgroup of archaea characterized by their ability to produce methane or other hydrocarbons as a primary metabolic product. Their unique features, including the lack of peptidoglycan-containing cell walls, distinct ribosomal features, and possession of membrane lipids, differentiated methanogens from bacteria (Madigan et al. 2018, Woese and Fox 1977). The outermost layer of archaea was typically composed of a protein layer known as the S-layer. This layer anchored to the cell wall through a para-crystalline arrangement of protein subunits. The S-layer served various functions, including protection from viruses, adhesion, cell shape definition, mechanical stability, and molecular sieving (van Wolferen et al. 2022).

3.3.2 Euryarchaeota

Euryarchaeota was a phylogenetically diverse group that encompassed methanogens. It was further divided into two classes: Class I and Class II. While initially grouped with Methanococcales in Class I, Methanopyrales and Methanobacteriales were later reclassified into Class II alongside Methanosarcinales (Luo et al. 2009). Based on 16S rRNA gene sequencing and other physiological characteristics, methanogens were classified into five orders: Methanobacteriales, Methanococcales, Methanomicrobiales, Methanosarcinales, and Methanopyrales (Liu 2010b, 2010a,

2010c, Liu and Whitman 2008). Each order exhibited distinct characteristics in terms of nutritional requirements, morphology, and substrate utilization.

3.3.3 Ecological and Biotechnological Significance

Archaea had significant ecological, biotechnological, and even human health implications due to their diversity. They played a central role in the evolutionary origins of eukaryotes and were involved in various biogeochemical processes. Advanced studies, enabled by culturing live cells from extreme environments, had contributed to a better understanding of archaeal biology. However, further research was needed to address the technical challenges associated with studying these organisms and to fully explore their potential applications (van Wolferen et al. 2022).

3.4 Evolutionary Relationships

Methanogenic archaea, or methanogens, are a diverse group of organisms capable of producing methane. They utilize three primary pathways for methanogenesis: CO_2 reduction, methyl-group reduction, and aceticlastic reaction. Based on their phylogeny and phenotypic characteristics, cultured methanogens are classified into five orders. However, many environments harbor uncultured methanogens that may represent new orders (Liu and Whitman 2008). Methanogenesis is a unique anaerobic respiration process that provides methanogens with the energy they need to grow. While no methanogens have been identified that can grow without producing methane, their methanogenic pathways are complex and require significant resources. Modern methanogens are thought to have an ancient ancestor within the Euryarchaeota (Sarmiento et al. 2011, Volmer et al. 2023). The increasing diversity of archaea on Earth offers insights into how their ancestors originated and diversified into extant forms. The ancient origin of methanogens has been widely accepted since the 1970s (Mei et al. 2023). This may be due to one of reasons, i.e., mutations, sunthrophy, or horizontal gene transfer. Syntrophy, a mutually beneficial interaction between microorganisms, allows microbial communities to survive in environments where individual organisms cannot. This phenomenon often involves the exchange of genes or molecules and is a common feature of microbial communities (Libby et al. 2019). Horizontal gene transfer, mediated by plasmids, is a significant driver of microbial evolution. Plasmids are extrachromosomal DNA elements that can replicate independently of the host chromosome. The genes encoded on plasmids are distinct from those affecting chromosomal genes and are subject to different genetic processes, such as genetic dominance, heteroplasmy, and segregational drift (Rodríguez-Beltrán et al. 2021). By introducing new genes into recipient bacteria, horizontal gene transfer can alter their genomes and lead to evolutionary changes (Zarlenga et al. 2022, Zarlenga et al. 2019).

3.5 Future Perspectives on Taxonomy and Evolution

Future research on biogas-producing microbiomes should focus on enhancing anaerobic digestion stability and efficiency. The continued advancement of

Omics-Technologies, particularly Metaproteomics and Metametabolomics, will be instrumental in achieving these goals. Metaproteomics can provide insights into the functional activities of microbial communities, allowing for the identification of key players in biogas production and the development of strategies to optimize their performance. Metametabolomics, on the other hand, can reveal the metabolic pathways involved in biogas generation and identify potential bottlenecks or limitations. By combining these approaches with traditional cultivation-based methods, researchers can gain a more comprehensive understanding of the microbial interactions and processes underlying anaerobic digestion, leading to the development of innovative and sustainable bioenergy solutions.

References

Abken, H.-J., Tietze, M., Brodersen, J., Bäumer, S., Beifuss, U. and Deppenmeier, U. (1998). Isolation and characterization of methanophenazine and function of phenazines in membrane-bound electron transport of *Methanosarcina mazei* Gö1. Journal of Bacteriology 180(8): 2027–2032. https://doi.org/10.1128/JB.180.8.2027-2032.1998.

Aparici-Carratalá, D., Esclapez, J., Bautista, V., Bonete, M.-J. and Camacho, M. (2023). Archaea: current and potential biotechnological applications. Research in Microbiology 174(7): 104080. https://doi.org/10.1016/j.resmic.2023.104080.

Ayesha, B. B., Gachinmath, S. and Sobia, C. (2022). Isolation of obligate anaerobes from clinical samples received for routine bacterial culture and sensitivity: a cross sectional study. Iranian Journal of Microbiology 14(2): 145–155. https://doi.org/10.18502/ijm.v14i2.9179.

Barker, H. A. (1936). Studies upon the methane-producing bacteria. Archiv Für Mikrobiologie 7(1–5): 420–438. https://doi.org/10.1007/BF00407414.

Dróżdż, M., Makuch, S., Cieniuch, G., Woźniak, M. and Ziółkowski, P. (2020). Obligate and facultative anaerobic bacteria in targeted cancer therapy: Current strategies and clinical applications. Life Sciences 261: 118296. https://doi.org/10.1016/j.lfs.2020.118296.

Ferraz Helene, L. C., Klepa, M. S. and Hungria, M. (2022). New insights into the taxonomy of bacteria in the genomic era and a case study with rhizobia. International Journal of Microbiology 2022: 4623713. https://doi.org/10.1155/2022/4623713.

Hatti-Kaul, R. and Mattiasson, B. (2016). Anaerobes in Industrial- and Environmental Biotechnology (pp. 1–33). https://doi.org/10.1007/10_2016_10.

Hoegenauer, C., Hammer, H. F., Mahnert, A. and Moissl-Eichinger, C. (2022). Methanogenic archaea in the human gastrointestinal tract. Nature Reviews Gastroenterology & Hepatology 19(12): 805–813. https://doi.org/10.1038/s41575-022-00673-z.

Janda, J. M. and Abbott, S. L. (2007). 16S rRNA gene sequencing for bacterial identification in the diagnostic laboratory: pluses, perils, and pitfalls. Journal of Clinical Microbiology 45(9): 2761–2764. https://doi.org/10.1128/JCM.01228-07.

Jones, L. S. and Howe, R. A. (2014). Microbial resistance and superbugs. pp. 257–285. *In*: Biofilms in Infection Prevention and Control. Elsevier. https://doi.org/10.1016/B978-0-12-397043-5.00015-3.

Jousimies-Somer, H. (1997). Recently described clinically important anaerobic bacteria: taxonomic aspects and update. Clinical Infectious Diseases : An Official Publication of the Infectious Diseases Society of America 25 Suppl 2: S78–87. https://doi.org/10.1086/516227.

Kim, M. and Chun, J. (2014). 16S rRNA Gene-Based Identification of Bacteria and Archaea using the EzTaxon Server (pp. 61–74). https://doi.org/10.1016/bs.mim.2014.08.001.

Kitahara, K. and Miyazaki, K. (2013). Revisiting bacterial phylogeny: Natural and experimental evidence for horizontal gene transfer of 16S rRNA. Mobile Genetic Elements 3(1): e24210. https://doi.org/10.4161/mge.24210.

Letzel, A.-C., Pidot, S. J. and Hertweck, C. (2013). A genomic approach to the cryptic secondary metabolome of the anaerobic world. Nat. Prod. Rep. 30(3): 392–428. https://doi.org/10.1039/C2NP20103H.

Li, J. S., Barber, C. C. and Zhang, W. (2019). Natural products from anaerobes. Journal of Industrial Microbiology and Biotechnology 46(3–4): 375–383. https://doi.org/10.1007/s10295-018-2086-5.

Libby, E., Hébert-Dufresne, L., Hosseini, S.-R. and Wagner, A. (2019). Syntrophy emerges spontaneously in complex metabolic systems. PLoS Computational Biology 15(7): e1007169. https://doi.org/10.1371/journal.pcbi.1007169.

Liu, Y. (2010a). Methanococcales. pp. 573–581. *In*: Handbook of Hydrocarbon and Lipid Microbiology. Springer Berlin Heidelberg. https://doi.org/10.1007/978-3-540-77587-4_44.

Liu, Y. (2010b). Methanopyrales. pp. 605–607. *In*: Handbook of Hydrocarbon and Lipid Microbiology. Springer Berlin Heidelberg. https://doi.org/10.1007/978-3-540-77587-4_47.

Liu, Y. (2010c). Taxonomy of methanogens. pp. 547–558. *In*: Handbook of Hydrocarbon and Lipid Microbiology. Springer Berlin Heidelberg. https://doi.org/10.1007/978-3-540-77587-4_42.

Liu, Y. and Whitman, W. B. (2008). Metabolic, phylogenetic, and ecological diversity of the methanogenic archaea. Annals of the New York Academy of Sciences 1125: 171–189. https://doi.org/10.1196/annals.1419.019.

Luo, H., Sun, Z., Arndt, W., Shi, J., Friedman, R. and Tang, J. (2009). Gene order phylogeny and the evolution of methanogens. PLoS ONE 4(6): e6069. https://doi.org/10.1371/journal.pone.0006069.

Madigan, M. T., Bender, K. S., Buckley, D. H., Sattley, W. M. and Stahl, D. A. (2018). Brock Biology of Microorganisms (15th ed.). Pearson.

Mei, R., Kaneko, M., Imachi, H. and Nobu, M. K. (2023). The origin and evolution of methanogenesis and Archaea are intertwined. PNAS Nexus 2(2). https://doi.org/10.1093/pnasnexus/pgad023.

Moore, E. R. B., Mihaylova, S. A., Vandamme, P., Krichevsky, M. I. and Dijkshoorn, L. (2010). Microbial systematics and taxonomy: relevance for a microbial commons. Research in Microbiology 161(6): 430–438. https://doi.org/10.1016/j.resmic.2010.05.007.

Noor, A. and Khetarpal, S. (2023). Anaerobic Infections. StatPearls Publishing.

Rodríguez-Beltrán, J., DelaFuente, J., León-Sampedro, R., MacLean, R. C. and San Millán, Á. (2021). Beyond horizontal gene transfer: the role of plasmids in bacterial evolution. Nature Reviews Microbiology 19(6): 347–359. https://doi.org/10.1038/s41579-020-00497-1.

Rupnik, M., Kovács, K. L., Nagaraja, T. G. and Allen-Vercoe, E. (2021). Anaerobes in the microbiome. Anaerobe 68: 102362. https://doi.org/10.1016/j.anaerobe.2021.102362.

Sarmiento, F. B., Leigh, J. A. and Whitman, W. B. (2011). Genetic Systems for Hydrogenotrophic Methanogens (pp. 43–73). https://doi.org/10.1016/B978-0-12-385112-3.00003-2.

Sebald, M. and Hauser, D. (1995). Pasteur, oxygen and the anaerobes revisited. Anaerobe 1(1): 11–16. https://doi.org/10.1016/S1075-9964(95)80353-X.

Sharma, S., Hashmi, M. and Valentino III, D. (2024). Actinomycosis. StatPearls Publishing.

Simon, J. (2013). Electron transport in facultative anaerobes. pp. 630–633. *In*: Encyclopedia of Biophysics. Springer Berlin Heidelberg. https://doi.org/10.1007/978-3-642-16712-6_32.

Sykes, J. E. (2014). Anaerobic bacterial infections. pp. 364–371. *In*: Canine and Feline Infectious Diseases. Elsevier. https://doi.org/10.1016/B978-1-4377-0795-3.00037-5.

van Wolferen, M., Pulschen, A. A., Baum, B., Gribaldo, S. and Albers, S.-V. (2022). The cell biology of archaea. Nature Microbiology 7(11): 1744–1755. https://doi.org/10.1038/s41564-022-01215-8.

Vítězová, M., Kohoutová, A., Vítěz, T., Hanišáková, N. and Kushkevych, I. (2020). Methanogenic microorganisms in industrial wastewater anaerobic treatment. Processes 8(12): 1546. https://doi.org/10.3390/pr8121546.

Volmer, J. G., McRae, H. and Morrison, M. (2023). The evolving role of methanogenic archaea in mammalian microbiomes. Frontiers in Microbiology 14: 1268451. https://doi.org/10.3389/fmicb.2023.1268451.

Wells, C. and Wilkins, T. (1996). Chapter 18. Clostridia: Sporeforming Anaerobic Bacilli. *In*: Baron, S. (ed.). Medical Microbiology. 4th edition. Galveston (TX): University of Texas Medical Branch at Galveston; 1996. University of Texas Medical Branch at Galveston.

Woese, C. R. and Fox, G. E. (1977). Phylogenetic structure of the prokaryotic domain: The primary kingdoms. Proceedings of the National Academy of Sciences 74(11): 5088–5090. https://doi.org/10.1073/pnas.74.11.5088.

Woese, C. R., Magrum, L. J. and Fox, G. E. (1978). Archaebacteria. Journal of Molecular Evolution 11(3): 245–252. https://doi.org/10.1007/BF01734485.

Zarlenga, D. S., Mitreva, M., Thompson, P., Tyagi, R., Tuo, W. and Hoberg, E. P. (2019). A tale of three kingdoms: members of the Phylum Nematoda independently acquired the detoxifying enzyme cyanase through horizontal gene transfer from plants and bacteria. Parasitology 146(4): 445–452. https://doi.org/10.1017/S0031182018001701.

Zarlenga, D., Thompson, P., Mitreva, M., Rosa, B. A. and Hoberg, E. (2022). Horizontal gene transfer provides insights into the deep evolutionary history and biology of Trichinella. Food and Waterborne Parasitology 27: e00155. https://doi.org/10.1016/j.fawpar.2022.e00155.

Chapter 4

Methane Production and Its Role in Global Carbon Cycling

Prasann Kumar[1,*] and *Joginder Singh*[2]

4.1 Introduction

A hydrocarbon compound known as methane, consisting of one carbon atom and four hydrogen atoms (CH_4), plays a significant role in the complex dynamics of Earth's climate system (Mostafa et al. 2023, Qi et al. 2022). The significance of carbon dioxide (CO_2) lies in its high potency as a greenhouse gas (GHG), which enables it to effectively trap heat in the atmosphere with significantly greater efficiency per molecule compared to CO_2. Although methane is less prevalent in the atmosphere compared to CO_2, its impact on global warming is significant, making it a central topic in the conversation about efforts to mitigate climate change. Comprehending the complexities of climate change requires a thorough understanding of the dynamics of methane production and its intricate role in the global carbon cycle. Methane emissions originate from various natural and human-induced sources, all contributing to the accumulation of methane in the atmosphere and its subsequent influence on the Earth's climate system. The sources of methane encompass a wide range of environments, including expansive wetlands and vibrant urban centres. Consequently, a comprehensive approach is required to investigate and address the diverse impacts of methane. This introductory section offers a concise summary of methane, its relevance in the dynamics of climate change, and the significance of investigating its generation and function within the global carbon cycle. By exploring these fundamental aspects, we establish the foundation for a more comprehensive analysis of the intricate nature of methane and its potential ramifications on our planet's future climate (Hadian et al. 2023, Omidkar et al. 2023,

[1] Department of Agronomy, School of Agriculture, Lovely Professional University, Phagwara, Punjab, 144411, India.
[2] Department of Botany, Nagaland University, Nagaland, India.
* Corresponding author: prasann0659@gmail.com

DOI: 10.1201/9781003506898-4

Tampio et al. 2024). The trajectory of methane in the Earth's atmosphere commences with its emission from diverse natural and human-induced origins. Upon emission, methane is introduced into the atmospheric greenhouse gas reservoir, thereby playing a role in the retention of outgoing longwave radiation, and subsequently leading to the warming of the Earth's surface. The significance of methane as a greenhouse gas is noteworthy, as it is estimated to have a Global Warming Potential (GWP) over 20 years that is approximately 84–87 times greater than that of carbon dioxide (CO_2). Nevertheless, this percentage decreases over extended periods due to methane's relatively brief existence in the atmosphere. Methane is derived from a wide range of natural sources, which include various ecological systems and processes. Wetlands, especially in tropical and boreal areas, are the primary natural origin of methane. They serve as habitats for microbial communities that can produce methane without needing oxygen, a process known as methanogenesis. Additional natural sources of methane include oceans, where methane hydrates have the potential to release methane into the water column and subsequently into the atmosphere. Additionally, terrestrial environments such as peatlands can generate methane through the anaerobic decomposition of organic matter. In conjunction with natural sources, human activities play a substantial role in generating methane emissions. Livestock farming, specifically agriculture, is a major contributor to methane emissions. Processes like enteric fermentation in ruminant animals and the breakdown of organic matter in manure management systems primarily cause this (Batgi and Dincer 2024, Tian et al. 2024, Zhou et al. 2023). The process of energy production, encompassing the extraction and distribution of fossil fuels, results in the emission of methane through leaks and venting during extraction procedures, as well as incomplete combustion during energy generation. In addition, human activities such as managing waste, such as waste disposal in landfills and treating wastewater, are major contributors to methane emissions. Landfills function as "methane factories" because organic waste undergoes anaerobic decomposition in environments lacking oxygen. Wastewater treatment facilities emit methane as a byproduct of microbial processes involved in sewage treatment. Methane is of utmost importance in the context of climate change. The abundance and long atmospheric lifetime of CO_2 make it the predominant driver of anthropogenic climate change. However, the potency of methane as a greenhouse gas enhances its influence on shorter timescales. The mitigation of near-term climate warming and the complementarity of efforts to reduce CO_2 emissions can be achieved through reductions in methane emissions. Hence, it is imperative to thoroughly comprehend methane generation, its origins, and its significance within the worldwide carbon cycle to develop productive strategies for mitigating climate change. The examination of methane generation and its function in the global carbon cycle carries substantial ramifications for scientific comprehension and policy development in tackling climate change. Methane emissions have the potential to initiate feedback loops that intensify the phenomenon of climate change. An example of this is the process of permafrost thawing in Arctic areas, which releases trapped methane, intensifying the warming effect. Comprehending these feedback mechanisms is imperative for precise predictions regarding future climate scenarios (Akimbekov et al. 2024, Cannone et al. 2024, Prabowo et al. 2024, Zhang et al. 2022). Implementing methane mitigation strategies can result in significant immediate climate advantages. Hence, policymakers necessitate precise data regarding methane emissions, origins, and patterns to devise efficacious policies aimed at diminishing emissions and achieving climate objectives delineated in global accords like the Paris Agreement. Enhancing our comprehension of methane production mechanisms, encompassing microbial pathways, biogeochemical cycling, and ecological interactions, augments the broader scientific understanding of Earth's biogeochemical cycles. This research improves our capacity to model and forecast forthcoming climate scenarios effectively. In contrast to carbon dioxide (CO_2), which exhibits a prolonged presence in the atmosphere spanning several centuries to millennia, methane possesses a comparatively brief atmospheric lifespan of approximately 12 years. Therefore, reducing methane emissions presents a more immediate prospect for mitigating the impacts of climate change. Examining methane production and determining efficient measures to reduce it can help exploit this potential. Methane emissions are frequently observed with other air pollutants, including Volatile Organic Compounds (VOCs) and Nitrogen Oxides (NOx). It is imperative to comprehend these interactions to evaluate the concurrent advantages of methane mitigation measures on air quality and human health. Examining methane production and its significance within the global carbon cycle is imperative to enhance scientific comprehension, guide policy-making, and alleviate the consequences of climate change. Through a comprehensive analysis of the intricate nature of methane emissions, it becomes possible to formulate precise approaches to mitigate emissions, mitigate climate-related hazards, and cultivate a more sustainable trajectory for forthcoming cohorts (Fig. 4.1).

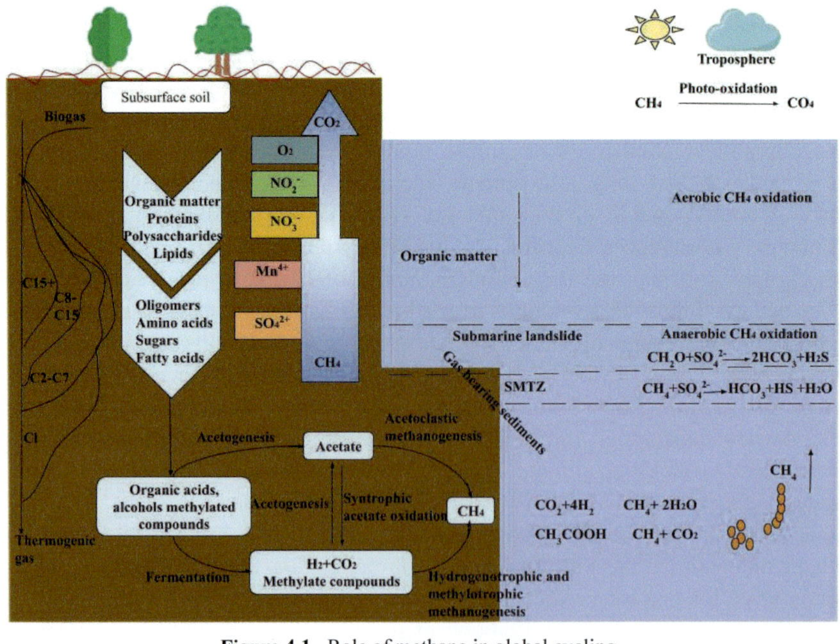

Figure 4.1. Role of methane in global cycling.

4.2 Methane Production Processes

Biogenic methane production pertains to methane synthesis via biological mechanisms, predominantly in habitats such as wetlands, oceans, and the gastrointestinal tracts of specific organisms. Methane emissions from natural sources have been present in the Earth's carbon cycle for millions of years. However, human activities have greatly modified these processes, increasing methane levels in the atmosphere. It is imperative to understand the underlying mechanisms that propel biogenic methane production and the various factors that impact emissions from natural sources to evaluate their contributions to climate change effectively. The largest natural source of biogenic methane emissions is wetlands, which encompass marshes, swamps, and bogs. The presence of saturated soils, anaerobic conditions, and abundant organic matter in these environments creates optimal conditions for the flourishing of methanogenic microorganisms. Organic matter gathers in saturated soils within wetland environments, where anaerobic bacteria break it down. The process is significantly influenced by methanogenic archaea, specialised microorganisms responsible for producing methane as a metabolic byproduct (García-Casado et al. 2024, Huang et al. 2023, Miao et al. 2022, Peng 2023, Peng et al. 2024, Thalasso et al. 2022). The process of organic matter decomposition involves the breakdown of intricate organic compounds, such as cellulose and lignin, by methanogens, resulting in methane emission into the Earth's atmosphere. Oceans are commonly acknowledged as reservoirs for methane due to microbial decomposition and physical mechanisms. However, specific marine habitats can also be the origins of biogenic methane emissions. Marine sediments harbour methane hydrates at elevated pressures and low temperatures, ice-like methane formations entrapped within a lattice of water molecules. Methane hydrates can dissociate and release methane into the water column and atmosphere when they are disturbed by geological or human-induced factors, such as underwater landslides or drilling activities. In marine sediments, microbial methane production occurs in oxygen-depleted zones, which are specific areas where anaerobic conditions promote the activity of methanogenic microorganisms. Termites, specifically those classified under the subfamily Macrotermitinae, contribute substantially to releasing methane through biological processes in tropical areas. Microbial symbionts in the termite gut assist in the breakdown of cellulose and other intricate carbohydrates, producing methane as a metabolic byproduct. Although termites contribute less to global methane emissions than other natural sources, their impact is significant in certain ecological groups, such as tropical forests and savannas. Many microbial processes play a significant role in generating biogenic methane in both terrestrial and aquatic environments. Methanogenic archaea employ diverse substrates, such as acetate, hydrogen, and carbon dioxide, under anaerobic conditions to generate methane via biochemical pathways, namely acetoclastic methanogenesis, hydrogenotrophic methanogenesis, and methylotrophic methanogenesis. A wide range of microbial processes can be observed in various habitats, encompassing the digestive systems of animals and sedimentary environments found in freshwater and marine ecosystems (Eisavi et al. 2024, Kadam and Yadav 2024, Litheko et al. 2023, Mao et al. 2024, Sabrekov et al. 2024, Sayyah et al. 2023, Sun et al. 2023). Environmental factors, including temperature, moisture, and substrate availability, significantly influence biogenic methane production. Warmer temperatures in wetland ecosystems promote microbial activity and expedite the decomposition of organic matter, resulting in increased methane emissions. Furthermore, the rates of wetland methane production can be influenced by various factors such as fluctuations in the water table, seasonal flooding, and vegetation composition. Variations in temperature and pressure influence methane hydrate stability in marine environments. Higher temperatures and lower water depths promote the release of methane. Organic substrates, including plant debris, residue, and animal waste, directly influence the methane production rates in biogenic environments. Wetlands that demonstrate elevated levels of primary productivity, which is characterised by a substantial presence of plant biomass, typically display increased methane emissions due to greater organic matter inputs. In aquatic environments, sediments abundant in nutrients provide a conducive environment for microbial communities that can metabolise organic compounds and generate methane as a byproduct of their metabolic processes. The composition and diversity of microbial communities heavily influence the regulation of biogenic methane production. Methanogenic archaea, the main methane producers in anaerobic environments, possess unique physiological and metabolic traits that impact their activity levels and the amount of methane they produce. Alterations in the structure of microbial communities, such as modifications in the composition of species or the prevalence of particular taxa, can potentially impact the dynamics of methane production in response to environmental disturbances or human-induced disruptions. Human activities have the potential to exert substantial pressure on the processes of biogenic methane production, resulting in modifications to methane emissions originating from natural sources. Alterations in land use, such as the removal of forests, the development of urban areas, and the expansion of agriculture, can disturb wetland habitats and expedite the breakdown of organic matter, leading to increased methane emissions. Furthermore, the runoff of nutrients resulting from agricultural activities has the potential to induce microbial activity within aquatic environments, thereby resulting in heightened methane production. The presence of human activities, such as offshore drilling and seabed mining, has the potential to disrupt marine environments, leading to the destabilisation of methane hydrates and subsequent release of methane into the water column (Akam et al. 2023, Hu et al. 2024, Lin et al. 2024, Minardi et al. 2023, Wang et al. 2024a–d). The processes of biogenic methane production are closely interconnected with the dynamics of climate change, resulting in potential feedback loops that contribute to the worsening of global warming. With the increase in temperatures, there is a potential for heightened microbial activity in wetlands and other biogenic environments, resulting in elevated levels of methane emissions. The process of permafrost thawing in Arctic regions has the potential to liberate substantial quantities of organic carbon, thereby stimulating microbial methane generation and exacerbating global warming. The feedback above of mechanisms highlights the interdependence between biogenic methane emissions and the wider consequences of climate change. Producing methane through biological processes, originating from natural sources such as wetlands, oceans, termites, and microbial processes, constitutes a substantial element within the Earth's carbon cycle. It is imperative to comprehend the various factors that impact biogenic methane emissions to effectively evaluate their contributions to atmospheric methane concentrations and their significance in the dynamics of climate change. By elucidating the mechanisms that drive biogenic methane production, researchers have the potential to enhance predictive models, provide valuable insights for policy decisions, and devise focused mitigation strategies to address this crucial aspect of the global methane budget effectively. Human activities make a substantial contribution to the overall release of methane on a global scale, increasing the amount of methane produced by natural sources and worsening the effects of climate change. Human-induced processes contribute to methane emission into the atmosphere through diverse channels, encompassing agricultural practices, extraction and combustion of fossil fuels, waste management, and industrial operations. Gaining insight into the mechanisms that drive human-caused methane production is crucial for devising efficient measures to decrease emissions and alleviate the consequences of climate change. The agricultural sector plays a substantial role in generating anthropogenic methane emissions, primarily through activities such as livestock production, rice cultivation, and manure management. The process of enteric fermentation, which occurs during the digestion of ruminant animals like cattle, sheep, and goats, is a notable contributor to the release of methane emissions. The rumen of these animals contains methanogenic microorganisms that break down ingested feed, especially fibrous plant materials, and produce methane as a byproduct. Methane emissions from enteric fermentation are influenced by various factors, such as the composition of the diet, the physiology of the animals involved, and the practices employed for management. Methods for reducing enteric methane emissions encompass dietary adjustments, using feed additives such as methane inhibitors, and implementing enhanced animal husbandry techniques (García-Luna and Ortiz 2024, Ma et al. 2024, Sayyah et al. 2022, Tan et al. 2024, Wang et al. 2024a,b, Zhang et al. 2023, Zhang et al. 2024b). An additional significant contributor to anthropogenic methane emissions is rice cultivation in flooded paddies. The presence of anaerobic conditions in waterlogged soils facilitates methane generation by microorganisms, primarily through organic matter decomposition and methanogenesis. Methane emissions originating from rice paddies predominantly arise from the generation of methane within the inundated soil and subsequent transportation of methane through plant tissues, commonly referred to as the “aerenchyma” pathway. Methane emissions from rice paddies can be mitigated by implementing management practices, such as Alternate Wetting and Drying (AWD) irrigation and mid-season drainage. These practices effectively modify soil redox conditions and restrict methanogenic activity. Livestock manure management is a major contributor to human-caused methane emissions. The anaerobic decomposition of manure in storage facilities, such as lagoons and pits, produces methane. When this methane is stored, handled, or applied to land, it has the potential to be released into the atmosphere. Various factors, including temperature, moisture content, and management practices such as anaerobic digestion and composting, influence methane emissions from manure management. Methane capture technologies, such as covered lagoons and anaerobic digesters, can reduce methane emissions from manure management facilities by capturing and recovering methane for energy generation. Human-caused methane emissions from the extraction and burning of fossil fuels occur at various stages, including production, processing, transportation, and use of fossil fuels like coal, oil, and natural gas. The primary sources of methane emissions in fossil fuel extraction are leaks and venting during drilling, extraction, and transportation activities. Fugitive methane emissions in oil and gas production arise due to equipment leaks, insufficient maintenance practices, and operational inefficiencies (Ghannadi and Dincer 2024, Pintér 2024, Vargas-Rodríguez et al. 2024, Xu et al. 2024, Y. Zhang et al. 2024). The extraction of natural gas from shale formations, commonly known as hydraulic fracturing (fracking), has the potential to emit methane into the atmosphere due to leakage from wellbores, equipment, and wastewater disposal sites. Enhancing Leak Detection And Repair (LDAR) programmes and implementing regulatory measures are imperative to mitigate methane emissions from fossil fuel extraction operations effectively. Anthropogenic methane emissions are also caused by the combustion of fossil fuels, such as coal, oil, and natural gas, for energy generation and transportation. The release of unburned methane into the atmosphere can result from incomplete combustion processes, particularly in engines and power plants. Methane emissions from fossil fuel combustion frequently coincide with the release of additional air pollutants, including nitrogen oxides (NOx) and volatile organic compounds (VOCs), intensifying the adverse effects on air quality and climate change. Implementing cleaner energy technologies, such as using renewable energy sources and adopting energy efficiency measures, has the potential to mitigate methane emissions linked to fossil fuel combustion. The anaerobic decomposition of organic matter in waste materials, resulting from waste management practices such as landfill disposal and wastewater treatment, substantially contributes to anthropogenic methane emissions. Landfills are one of the primary human-caused sources of methane emissions on a global scale. Microbial communities engage in the anaerobic decomposition of organic waste that is deposited in landfills, resulting in the production of methane as a metabolic byproduct. Methane emissions originating from landfills can be released into the atmosphere via surface emissions, landfill gas vents, and leachate collection systems. Methane emissions from landfills are affected by the composition of the waste, the moisture content, and the practices used to manage the landfill. Methane capture and utilisation systems, such as landfill gas collection and flaring, can capture methane for energy production or its conversion into less potent greenhouse gases to mitigate emissions. Anthropogenic methane emissions are primarily caused by the anaerobic decomposition of organic matter in sewage sludge, which occurs in wastewater treatment facilities (Bidart et al. 2022, He et al. 2023, Ji and Shen 2024, Keller et al. 2023, Oni et al. 2023, Su et al. 2022, L. Wang et al. 2024). Microbial communities engage in the metabolic breakdown of organic compounds in sewage sludge during wastewater treatment processes, such as anaerobic digestion. This metabolic process results in the production of methane as a byproduct. Wastewater treatment plants can release methane during sludge storage, handling, and disposal.

Methane capture and utilisation technologies, such as anaerobic digesters and biogas recovery systems, can capture methane emissions from wastewater treatment facilities. This capture process enables the generation of energy or heat, reducing greenhouse gas emissions. Anthropogenic methane emissions are generated by industrial activities, encompassing chemical production, manufacturing, and waste incineration through various processes and byproducts. Methane can be released as a byproduct or fugitive emissions during chemical manufacturing processes, especially organic compounds and fossil fuel feedstocks. Methane emissions can arise from chemical synthesis, purification, and storage processes, as well as from equipment and pipeline leaks and venting. Reducing methane emissions from chemical manufacturing processes can be achieved by implementing cleaner production technologies, including process optimisation and emissions control measures. Methane emissions can be generated through combustion, chemical reactions, and waste disposal practices in various industrial manufacturing processes, such as metal production, cement manufacturing, and food processing. As an illustration, the production processes of iron and steel that utilise coal or coke as a reducing agent have the potential to emit methane as a byproduct resulting from incomplete combustion. Likewise, food processing facilities have the potential to generate methane emissions due to the decomposition of organic waste or the anaerobic digestion processes (Akimbekov et al. 2024, Batgi and Dincer 2024, Hadian et al. 2023, Prabowo et al. 2024, Thalasso et al. 2022, Tian et al. 2024, Zhang et al. 2022, Zhou et al. 2023). The mitigation of methane emissions from industrial manufacturing processes can be facilitated by implementing energy-efficient technologies and adopting effective waste management practices. Methane can be emitted as a byproduct of combustion processes in waste incineration facilities, which are utilised to dispose of municipal solid waste and hazardous materials. Methane and other air pollutants, including carbon monoxide (CO) and nitrogen oxides (NOx), can be generated through the incomplete combustion of organic waste materials. The mitigation of methane emissions from waste incineration can be achieved by implementing sophisticated technologies, including high-temperature combustion and flue gas treatment systems. These technologies facilitate the achievement of complete combustion and the regulation of emissions. Global methane emissions are significantly influenced by anthropogenic methane production resulting from agricultural activities, fossil fuel extraction and combustion, waste management, and industrial processes. Addressing human-caused methane emissions necessitates collaborative endeavours spanning various industries, encompassing cleaner technologies, enhanced waste management strategies, and regulatory interventions to diminish fugitive emissions and foster methane capture and utilisation. To mitigate the impacts of climate change and transition towards a more sustainable and resilient future, society should focus on addressing human-caused sources of methane emissions (Fig. 4.2).

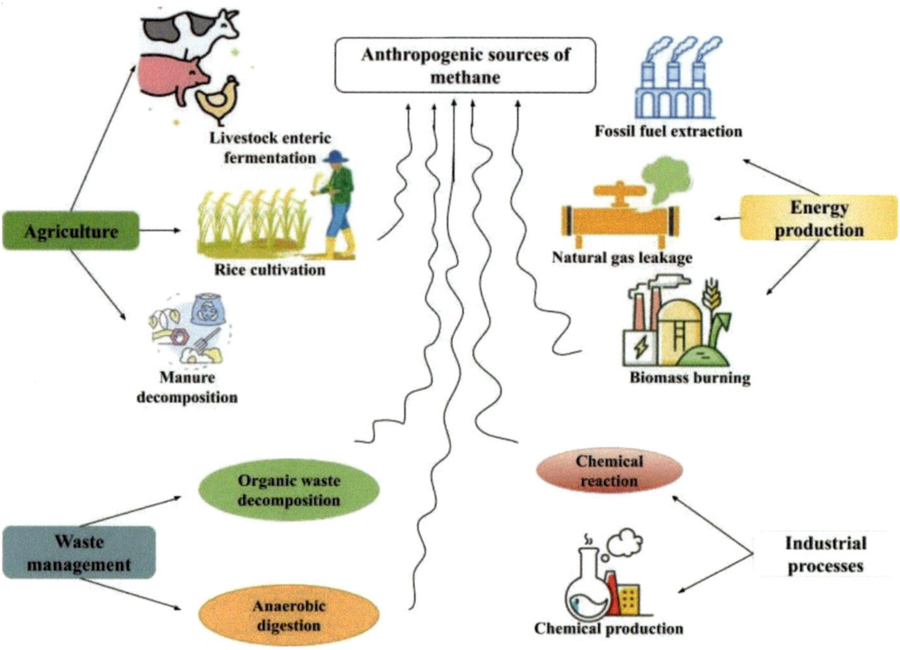

Figure 4.2. Anthropogenic cause of methane production.

4.3 Methane's Role in the Carbon Cycle

Methane, an influential greenhouse gas (GHG), is pivotal in the Earth's carbon cycle, exerting its impact on climate change dynamics through its interactions with the atmosphere, biosphere, and geosphere. Gaining insight into the function of methane in the carbon cycle is crucial for understanding its influence on the Earth's climate and developing efficient approaches to alleviate climate change (Akam et al. 2023, García-Luna and Ortiz 2024, Hu et al. 2024, Lin et al. 2024, Ma et al. 2024, Minardi et al. 2023, Sabrekov et al. 2024, Sayyah et al. 2022, X. Wang et al. 2024, J. Wang et al. 2024, Zhang et al. 2023, M. Zhang et al. 2024). Methane is a highly potent greenhouse gas that plays a significant role in the greenhouse effect, an essential natural mechanism responsible for regulating the Earth's temperature to a level suitable for sustaining life. Like carbon dioxide (CO_2) and other greenhouse gases (GHGs), methane can absorb and release infrared radiation. This allows it to retain heat in the Earth's atmosphere and cause its surface to warm. The efficacy of methane as a greenhouse gas stems from its molecular composition, which enables it to more efficiently absorb and re-emit infrared radiation compared to CO_2 at the molecular level. Radiative forcing is a measure that quantifies the disparity between the amount of solar radiation that the Earth absorbs and the amount of thermal radiation that is emitted into space. An increase in greenhouse gas (GHG) concentrations leads to positive radiative forcing, which intensifies the greenhouse effect and causes the Earth's surface to warm. The Global Warming Potential (GWP) of methane, which quantifies its capacity to trap heat compared to CO_2 within a defined time frame, is notably greater than that of CO_2 when considering individual molecules. Despite having a shorter duration in the atmosphere than CO_2, methane's greater global warming potential (GWP) significantly contributes to its substantial influence on climate change, especially over shorter periods. The interaction between methane and carbon cycle components, such as the atmosphere, biosphere, and geosphere, significantly impacts carbon fluxes and feedback mechanisms (García-Luna and Ortiz 2024, Ghannadi and Dincer 2024, He et al. 2023, Ma et al. 2024, Sayyah et al. 2022, Tan et al. 2024, C. Wang et al. 2024, L. Wang et al. 2024, Xu et al. 2024, Zhang et al. 2023, M. Zhang et al. 2024). Methane exhibits chemical reactions with hydroxyl radicals (OH) in the atmosphere, resulting in oxidation and subsequent elimination from the atmosphere. The hydroxyl radical plays a crucial role as an atmospheric oxidant by eliminating methane and other minor atmospheric pollutants. Nevertheless, the atmospheric lifespan of methane, estimated to be around 9–12 years, enables it to endure in the atmosphere for a considerable duration, thereby exerting substantial radiative forcing effects. Many ecological systems, such as wetlands, terrestrial soils, oceans, and animal digestive systems, produce and consume methane. Methanogenic microorganisms in anaerobic environments, such as wetlands and rice paddies, engage in microbial processes, namely acetoclastic methanogenesis and hydrogenotrophic methanogenesis, to generate methane. In contrast, methanotrophic bacteria found in highland soils and aerobic environments can utilise methane as a carbon and energy source, reducing its emission into the atmosphere. The equilibrium between methane generation and utilisation in the biosphere impacts atmospheric methane levels and plays a role in feedback mechanisms that govern climate. Marine sediments and permafrost regions contain methane hydrates, which are ice-like structures that contain methane trapped within a lattice of water molecules. Methane hydrates are extensive methane repositories stored beneath the Earth's surface. If these hydrates become destabilised due to fluctuations in temperature, pressure, or ocean currents, they have the potential to release significant amounts of methane into the atmosphere. The phenomenon of permafrost thawing in Arctic regions, a consequence of climate change, presents a substantial hazard of releasing sequestered methane into the atmosphere, thereby exacerbating the effects of global warming and climate change. The involvement of methane in the carbon cycle plays a significant role in feedback mechanisms that can potentially exacerbate climate change through its impact on the Earth's radiative balance and biogeochemical cycles. Permafrost thawing in Arctic regions releases stored organic carbon, such as methane, into the atmosphere, exacerbating global warming. The acceleration of microbial decomposition of organic matter resulting from the thawing of permafrost contributes to heightened methane emissions and subsequent Arctic warming. The permafrost carbon feedback, a positive feedback loop, intensifies the effects of climate change and presents substantial obstacles to climate mitigation endeavours (Akimbekov et al. 2024, Batgi and Dincer 2024, Cannone et al. 2024, Huang et al. 2023, Omidkar et al. 2023, Peng 2023, Peng et al. 2024, Prabowo et al. 2024, Thalasso et al. 2022, Tian et al. 2024, Zhang et al. 2022, Zhou et al. 2023). The presence of methane hydrates in the Earth's climate system introduces an additional possibility for positive feedback. Methane hydrates can be destabilised by alterations in temperature, pressure, or ocean currents, releasing methane into the atmosphere and exacerbating global warming. The precise magnitude and timing of methane hydrate release are still not fully understood. However, the possibility of significant and swift methane emissions resulting from the destabilisation of hydrates highlights the significance of comprehending and observing methane dynamics in marine sediments. Alterations can influence methane emissions and feedback processes in the carbon cycle in land use and management practices. Deforestation, agricultural expansion, and wetland drainage can modify methane fluxes by altering the ecosystem's characteristics and the microorganisms' activity. Farming activities such as cultivating rice and raising livestock also contribute to the release of methane, thereby worsening the effects of climate change. Methane plays a significant role in the carbon cycle, extending beyond its classification as a potent greenhouse gas. It affects carbon fluxes and feedback processes by interacting with the atmosphere, biosphere, and geosphere. Researchers can develop targeted mitigation strategies to address anthropogenic emissions and mitigate climate change impacts by comprehending the contributions of methane to climate change dynamics and feedback mechanisms that amplify global warming. It is crucial to decrease methane emissions and encourage carbon storage in both natural and managed environments to achieve climate stabilisation objectives and guarantee a sustainable future for future generations (Fig. 4.3).

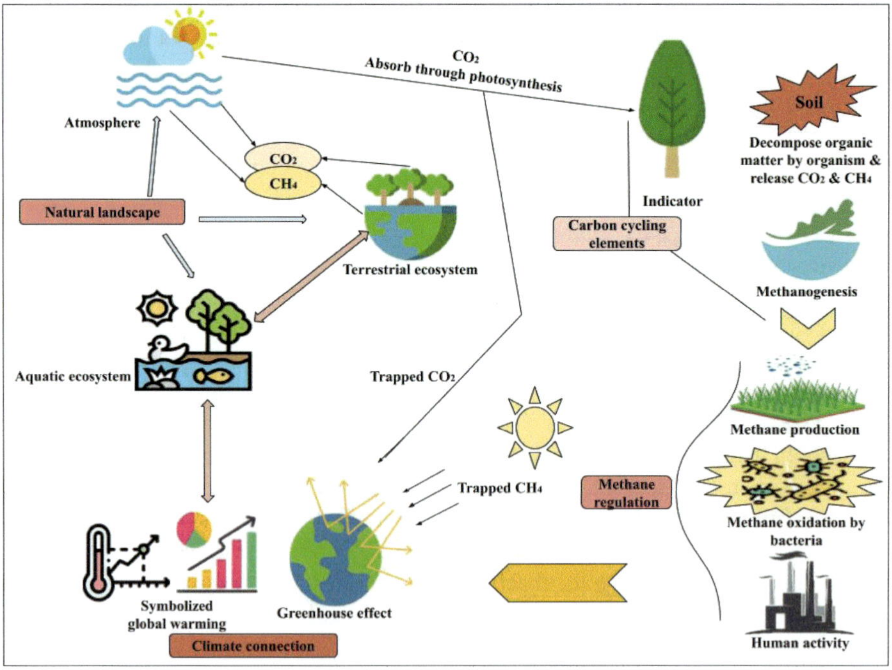

Figure 4.3. Carbon cycling and its regulation by the production of methane.

4.4 Impact on Climate Change

Methane's contribution to climate change is significant, with implications for global warming, regional and localised effects, and climate mitigation efforts. Understanding the relative contributions of methane compared to carbon dioxide (CO_2) emissions and the regional and local impacts of methane emissions is crucial for developing effective strategies to mitigate climate change. Methane is a potent greenhouse gas, second only to CO_2 in terms of its contribution to global warming. While CO_2 is more abundant and has a longer atmospheric lifetime, methane has a higher global warming potential (GWP) per molecule (Ghannadi and Dincer 2024, He et al. 2023, Ma et al. 2024, Tan et al. 2024, Vargas-Rodríguez et al. 2024, C. Wang et al. 2024, L. Wang et al. 2024, Xu et al. 2024, M. Zhang et al. 2024, Y. Zhang et al. 2024). Over a 20-year timeframe, methane's GWP is estimated to be approximately 84–87 times greater than that of CO_2, although this figure diminishes over longer timescales due to methane's shorter atmospheric lifetime. Despite its lower atmospheric concentration than CO_2, methane's higher radiative forcing potential significantly drives near-term climate warming. Anthropogenic methane emissions contribute approximately 25 to 30% of total anthropogenic radiative forcing since pre-industrial times, making them a critical factor in understanding and addressing climate change. While methane emissions are less abundant than CO_2 emissions, their impact on global warming per unit of mass is substantially more significant. This difference in radiative forcing efficiency underscores the importance of addressing methane emissions alongside CO_2 emissions in climate mitigation efforts. Additionally, methane and CO_2 emissions often originate from different sources and sectors. CO_2 emissions primarily result from fossil fuel combustion, industrial processes, and land use changes such as deforestation. In contrast, methane emissions arise from natural sources such as wetlands and agricultural activities and anthropogenic sources, including livestock production, waste management, and fossil fuel extraction. Addressing methane emissions offers a unique opportunity to achieve significant climate benefits in the short term due to methane's shorter atmospheric lifetime compared to CO_2. Reductions in methane emissions can help mitigate near-term climate warming, while complementary efforts to reduce CO_2 emissions focus on longer-term climate stabilisation goals. Methane emissions can have diverse regional and localised effects on climate, ecosystems, and human health, depending on the sources and distribution of emissions. Methane's role in climate change extends beyond its direct radiative forcing effects to include feedback mechanisms that can amplify warming (Akam et al. 2023, Bidart et al. 2022, García-Luna and Ortiz 2024, Ghannadi and Dincer 2024, He et al. 2023, Ma et al. 2024, Pintér 2024, Sayyah et al. 2022, Tan et al. 2024, C. Wang et al. 2024, Y. Zhang et al. 2024, Zhang et al. 2023, M. Zhang et al. 2024). For example, thawing permafrost in Arctic regions releases methane and carbon dioxide stored in frozen soils, contributing to further warming and degradation in a positive feedback loop. Similarly, methane emissions from wetlands, rice paddies, and other ecosystems can exacerbate climate change impacts by altering regional temperature patterns, precipitation regimes, and atmospheric composition. Methane emissions can influence ecosystem dynamics and biodiversity, particularly in wetland and aquatic environments. Increased methane concentrations in freshwater and marine ecosystems can alter nutrient cycling, aquatic productivity, and species composition, with cascading effects on food webs and ecosystem services. Methane emissions from terrestrial ecosystems can also affect soil carbon dynamics, microbial communities, and vegetation composition, influencing ecosystem resilience to climate change and disturbances. Methane emissions can indirectly impact human health by contributing to climate change and air pollution. Elevated methane concentrations in the atmosphere contribute to global warming, extreme weather events, and sea-level rise, with implications for human health and well-being. Additionally, methane emissions often co-occur with other air pollutants such as volatile organic compounds (VOCs), nitrogen oxides (NOx), and particulate matter, exacerbating respiratory illnesses, cardiovascular diseases, and other health conditions. As the Paris Agreement outlines, addressing methane emissions is essential for achieving climate stabilisation goals and limiting global warming to below 2 degrees Celsius above pre-industrial levels. Methane mitigation offers several benefits for climate change mitigation efforts, including reducing methane emissions, which can yield significant near-term climate benefits due to methane's shorter atmospheric lifetime than CO_2. Methane mitigation measures, such as improved waste management, methane capture and utilisation, and agricultural practices, can help mitigate near-term climate warming and complement efforts to reduce CO_2 emissions from fossil fuel combustion and deforestation. Many methane mitigation strategies also offer co-benefits for air quality and human health by reducing co-pollution emissions such as VOCs, NOx, and particulate matter. For example, capturing methane from landfills and wastewater treatment facilities reduces greenhouse gas emissions and prevents the release of volatile organic compounds and odorous compounds associated with waste decomposition. Addressing methane emissions requires a combination of technological innovations, policy interventions, and international cooperation (Akimbekov et al. 2024, Batgi and Dincer 2024, Cannone et al. 2024, García-Casado et al. 2024, Huang et al. 2023, Kadam and Yadav 2024, Litheko et al. 2023, Miao et al. 2022, Peng 2023, Peng et al. 2024, Prabowo et al. 2024, Tampio et al. 2024, Thalasso et al. 2022, Tian et al. 2024, Zhang et al. 2022). Methane mitigation technologies, such as methane capture and storage, anaerobic digestion, and biogas recovery, offer opportunities to reduce emissions across multiple sectors, including agriculture, waste management, and energy production. Additionally, policy measures such as regulations, incentives, and carbon pricing mechanisms can incentivise emissions reductions and promote the adoption of cleaner technologies and practices. Methane's impact on climate change extends beyond its role as a greenhouse gas, including regional and localised effects on ecosystems, human health, and socio-economic systems. Addressing methane emissions is essential for achieving climate stabilisation goals and limiting global warming, with significant implications for near-term climate benefits, air quality, and sustainable development. By prioritising methane mitigation efforts alongside CO_2 emissions reductions, society can mitigate climate change impacts and transition towards a more sustainable and resilient future. Methane emissions exhibit spatial and temporal variability, with specific regions and sectors serving as hotspots for emissions due to natural and anthropogenic factors. Natural sources such as wetlands, permafrost regions, and geological seeps contribute to methane emissions. Tropical wetlands, including the Amazon Basin and Southeast Asian peatlands, are among the most significant natural sources of methane due to their high rates of organic matter decomposition and methanogenesis. Additionally, thawing permafrost in Arctic regions releases stored methane and carbon dioxide, amplifying climate change impacts and contributing to positive feedback mechanisms. Anthropogenic activities such as agriculture, fossil fuel production, and waste management also contribute to methane emissions in specific regions and sectors (Akam et al. 2023, García-Luna and Ortiz 2024, Ghannadi and Dincer 2024, He et al. 2023, Ma et al. 2024, Pintér 2024, Sayyah et al. 2022, Tan et al. 2024, C. Wang et al. 2024, Y. Zhang et al. 2024, Zhang et al. 2023, M. Zhang et al. 2024). Agricultural hotspots include regions with intensive livestock production, rice cultivation, and manure management, such as South Asia, China, and North America. Fossil fuel extraction and processing facilities, including oil and gas operations, coal mines, and landfills, are concentrated in some geographic regions, leading to localised emissions hotspots. Urban areas with dense populations and industrial activities may also exhibit elevated methane concentrations due to emissions from transportation, waste disposal, and energy consumption. Methane emissions are subject to uncertainties stemming from measurement techniques, emission factors, and data availability, particularly in regions with limited monitoring infrastructure and data collection efforts. Methane leakage from infrastructure such as pipelines, wells, and storage facilities can contribute to emissions uncertainties in the fossil fuel industry. Incomplete combustion, fugitive emissions, and equipment leaks during fossil fuel extraction, processing, and transportation can produce unaccounted methane emissions, contributing to atmospheric concentrations. Improving leak detection and repair (LDAR) programs, implementing emission reduction technologies, and enhancing transparency and reporting standards can help reduce uncertainties associated with methane leakage from fossil fuel infrastructure. Methane emissions from agricultural activities, particularly enteric fermentation, and manure management, are subject to uncertainties due to variations in livestock populations, diet composition, and management practices. Estimating methane emissions from agricultural sources requires accurate emission factors, activity data, and spatial distribution information, which may vary across regions and livestock systems. Improving data collection methods, developing region-specific emission factors, and implementing mitigation measures can help reduce uncertainties associated with agricultural methane emissions. Methane mitigation efforts offer several co-benefits beyond climate change mitigation, including improvements in air quality, public health, and sustainable development. Reducing methane emissions can improve air quality by lowering co-pollutant concentrations such as volatile organic compounds (VOCs), nitrogen oxides (NOx), and particulate matter. Methane mitigation measures such as methane capture and utilisation, landfill gas recovery, and anaerobic digestion can help reduce co-pollutant emissions associated with waste decomposition and combustion processes, improving local air quality and public health. Improving air quality through methane mitigation can have significant health benefits for populations exposed to air pollution (Eisavi et al. 2024, Kadam and Yadav 2024,

Litheko et al. 2023, Mao et al. 2024, Sabrekov et al. 2024, Sayyah et al. 2023, Sun et al. 2023, X. Wang et al. 2024). Reducing emissions of delicate Particulate Matter (PM2.5), Ozone (O3), and other air pollutants associated with methane emissions can decrease the incidence of respiratory illnesses, cardiovascular diseases, and other health conditions. Additionally, reducing exposure to air pollutants can improve affected communities' quality of life, productivity, and well-being, particularly in urban areas with high pollution levels. Methane mitigation aligns with multiple Sustainable Development Goals (SDGs), including climate action, clean energy, sustainable cities, and responsible consumption and production. By reducing methane emissions and promoting sustainable practices in agriculture, waste management, and energy production, society can advance towards achieving SDGs related to poverty alleviation, food security, environmental conservation, and social equity. Methane's impact on climate change extends beyond its role as a greenhouse gas to include regional and localised effects, uncertainties in emissions estimation, and co-benefits of mitigation efforts. Addressing methane emissions requires comprehensive strategies that target natural and anthropogenic sources, improve monitoring and measurement techniques, and prioritise co-benefits such as air quality improvement and public health (Fig. 4.4). By integrating methane mitigation efforts into broader climate mitigation and sustainable development agendas, society can mitigate climate change impacts and build a more resilient and equitable future for all.

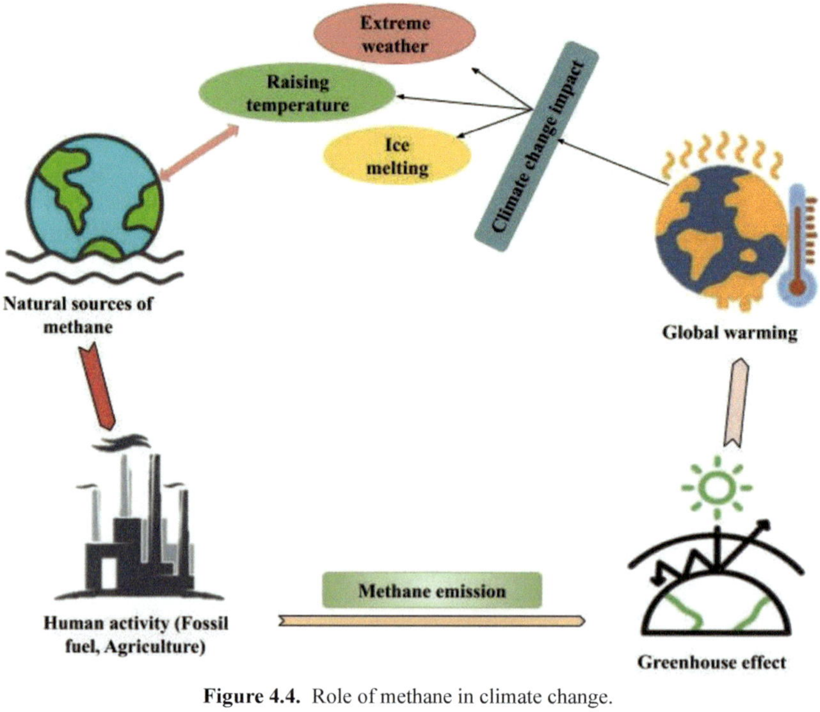

Figure 4.4. Role of methane in climate change.

4.5 Monitoring and Measurement Techniques

Precise measurement of methane emissions is crucial for comprehending its impact on climate change and guiding efficient measures to reduce it. A variety of methodologies are employed in the field of monitoring and measurement, which encompass remote sensing techniques as well as ground-based monitoring networks. Nevertheless, there are obstacles and constraints in precisely measuring methane emissions, underscoring the necessity for ongoing advancements and enhancements in measurement technologies. Remote sensing techniques present invaluable tools for monitoring methane emissions across expansive spatial scales and varied landscapes. These methods employ satellite-based sensors, aerial surveys, and ground-based instruments to detect and quantify atmospheric methane concentrations (Akam et al. 2023, García-Luna and Ortiz 2024, Ma et al. 2024, Oni et al. 2023, Pintér 2024, Tan et al. 2024, C. Wang et al. 2024, Xu et al. 2024, M. Zhang et al. 2024, Zhang et al. 2023). The Tropospheric Monitoring Instrument (TROPOMI) and the Greenhouse Gases Observing Satellite (GOSAT) are satellite-based remote sensing platforms that offer worldwide coverage and precise spatial resolution for monitoring atmospheric methane concentrations. The satellites employ spectrometers to quantify the amount of sunlight reflected or absorbed by methane molecules in the Earth's atmosphere. This enables maps to measure methane concentration and estimate methane emissions. Satellite-based remote sensing offers several benefits, such as extensive coverage, uninterrupted time, and the capability to monitor distant or hard-to-reach areas. Nevertheless, the quality and accuracy of data can be compromised by various factors, including cloud cover, surface reflectance, and atmospheric interference. This is especially true in regions characterised by intricate terrain or atmospheric conditions. UAVs equipped with remote sensing instruments, such as spectrometers or gas analysers, provide enhanced spatial resolution and increased flexibility compared to satellite-based platforms when conducting aerial surveys. Aerial surveys can selectively survey particular regions or sources of significance, such as oil and gas facilities, landfills, or agricultural areas, to obtain comprehensive measurements of methane concentration and emissions estimates. Advanced imaging techniques, such as thermal infrared cameras or hyperspectral sensors, may be employed in these surveys to enhance the sensitivity of methane plumes and source detection. Aerial surveys face challenges such as high operational expenses, logistical limitations, and restricted coverage compared to satellite-based observations. Uncrewed Aerial Vehicles (UAVs), commonly called drones, provide a versatile and economically viable means of conducting targeted methane measurements in regions inaccessible by human-operated aircraft or terrestrial instrument (Akimbekov et al. 2024, García-Casado et al. 2024, Hadian et al. 2023, Huang et al. 2023, Kadam and Yadav 2024, Litheko et al. 2023, Miao et al. 2022, Omidkar et al. 2023, Peng 2023, Peng et al. 2024, Thalasso et al. 2022, Tian et al. 2024, Zhang et al. 2022). Uncrewed Aerial Vehicles (UAVs) equipped with methane sensors or gas analysers can gather precise spatial data about specific areas, such as agricultural fields, wetlands, or industrial facilities. This data can be utilised to identify and quantify the sources of methane emissions. The benefits of using UAVs for measurements include quick deployment, the ability to customise flight paths, and the ability to make observations at low altitudes, which improves the sensitivity and accuracy of detection. Nevertheless, using uncrewed aerial vehicles (UAVs) for extensive methane monitoring may be hindered by various obstacles, including but not limited to restricted flight endurance, limited payload capacity, and regulatory limitations. Methane concentrations in the atmosphere are quantified by ground-based monitoring networks, which employ stationary or mobile measurement stations equipped with methane sensors, analysers, or spectroscopic instruments. The networks offer comprehensive spatial and temporal information regarding methane emissions from diverse sources and environments. Fixed monitoring stations are strategically positioned at various locations, including urban centres, industrial sites, and natural gas infrastructure. These stations employ gas chromatographs, Cavity Ring-Down Spectrometers (CRDS), or Fourier transform infrared (FTIR) spectrometers to measure atmospheric methane concentrations continuously. These stations offer extensive datasets for monitoring the patterns, fluctuations, and spatial variations in methane emissions over an extended period. Fixed monitoring stations offer several benefits, such as high-quality data, consistency, and comparability. These attributes greatly facilitate the analysis of trends and the quantification of emissions. Nevertheless, the extrapolation of station data to broader geographic scales may be influenced by limitations such as spatial representativeness and site-specific biases. Methane atmospheric concentrations can be measured flexibly and targeted using mobile monitoring platforms encompassing vehicles, aircraft, and handheld instruments. Mobile platforms equipped with methane sensors or analysers can navigate various landscapes, urban environments, and industrial establishments to detect and analyse sources of methane emissions (Akam et al. 2023, García-Luna and Ortiz 2024, Ghannadi and Dincer 2024, Ma et al. 2024, Pintér 2024, Sayyah et al. 2022, Tan et al. 2024, Vargas-Rodríguez et al. 2024, C. Wang et al. 2024, Xu et al. 2024, Y. Zhang et al. 2024, Zhang et al. 2023, M. Zhang et al. 2024). Mobile measurements provide several benefits, including the ability to be deployed quickly, provide spatial coverage, and survey expansive areas or regions that are difficult to access. Nevertheless, the accuracy and reliability of mobile measurements can be influenced by various challenges, including instrument calibration, data interpretation, and logistical constraints. Although there have been improvements in monitoring and measurement methods, multiple obstacles and restrictions exist in accurately measuring methane emissions, especially on a regional and global level. Natural processes, human activities, and atmospheric dynamics influence spatial and temporal variability of methane emissions. Methane concentrations and emission estimation efforts can be complicated by variations in emission sources, meteorological conditions, and atmospheric transport. The challenge of monitoring networks and measurement campaigns lies in attaining representative and comprehensive coverage of methane emissions across various landscapes, ecosystems, and sectors. Identifying and assigning methane emissions to particular sources, such as natural gas production, agriculture, or landfills, is crucial to comprehend emission patterns and guide strategies for reducing them. Nevertheless, differentiating emission sources solely through atmospheric measurements can present difficulties, especially in areas where multiple sources coexist, or emission pathways are intricate. The source attribution and differentiation capabilities can be enhanced by integrating atmospheric measurements with complementary data sources, including inventory data, isotopic analysis, and modelling simulations. Accurate and reliable measurements of atmospheric concentrations are necessary for quantifying methane emissions. However, these measurements are susceptible to uncertainties arising from instrument calibration, sampling techniques, and data processing algorithms. The process of validating measurement results through independent datasets, reference standards, and intercomparison exercises is crucial to evaluating the accuracy of measurements, detecting any biases, and enhancing the data quality (Akimbekov et al. 2024, Cannone et al. 2024, García-Casado et al. 2024, Huang et al. 2023, Kadam and Yadav 2024, Litheko et al. 2023, Miao et al. 2022, Peng 2023, Peng et al. 2024, Tampio et al. 2024, Tian et al. 2024). The mitigation of measurement uncertainties necessitates the implementation of continuous calibration and quality assurance protocols, as well as the establishment of collaborative initiatives among research institutions, government agencies, and industry stakeholders. Data harmonisation, interoperability, and synthesis challenges arise when integrating and synthesising methane measurement data from various sources and platforms, such as remote sensing observations, ground-based monitoring networks, and modelling simulations. Standardised protocols, data formats, metadata standards, interoperable data-sharing platforms, and analytical tools are necessary for smooth data integration and synthesis across spatial and temporal scales. Promoting stakeholder data sharing, collaboration, and knowledge exchange is a crucial objective of collaborative initiatives such as the Global Methane Initiative (GMI) and the Methane Remote Sensing Consortium (MERLIN) established by the European Space Agency. Measurement and monitoring techniques are crucial in quantifying methane emissions and providing valuable insights for climate mitigation endeavours. Remote sensing methods, ground-based monitoring networks, and mobile measurement platforms offer complementary approaches for detecting, quantifying, and attributing methane sources and emissions. Tackling obstacles such as spatial variability, source attribution, measurement uncertainty, and data integration necessitates collaboration across different fields, technological advancements, and ongoing enhancements in measurement technologies. By deepening our comprehension of methane emissions and their underlying causes, society can formulate precise measures to mitigate methane's impact on climate change and attain climate stabilisation objectives (Table 4.1).

Table 4.1. Different types of monitoring and measurement techniques (Source: Batgi and Dincer 2024, Cannone et al. 2024, Tampio et al. 2024, Tian et al. 2024, Zhang et al. 2022).

Technique	Description
1. Remote Sensing	It utilises satellite-based sensors to measure methane concentrations in the atmosphere by analysing spectral signatures absorbed or emitted by methane molecules.
2. Ground-based Sensors	It is deployed on the ground to measure methane concentrations directly. It utilises various technologies, such as infrared absorption, laser-based spectroscopy, and gas chromatography.
3. Aircraft-based Measurements	It involves mounting sensors on aircraft to conduct airborne surveys, allowing for the measurement of methane concentrations at different altitudes and across larger geographical areas.
4. Mobile Laboratory	Mobile units equipped with methane sensors for on-site measurements. It can be deployed to conduct surveys in areas where fixed monitoring stations are not available or feasible.
5. Flux Chambers	They are placed over soil or water surfaces to capture methane emissions. The collected gases are then analysed to determine emission rates and patterns.
6. Eddy Covariance	Measures the methane exchange between the Earth's surface and the atmosphere by analysing fluctuations in wind speed and methane concentrations within the planetary boundary layer.
7. Soil Gas Sampling	It involves collecting gas samples from soil depths using probes or wells. The collected gases are then analysed to determine methane concentrations, providing insights into soil methane fluxes.
8. Isotopic Analysis	Measures the isotopic composition of methane molecules to distinguish between biogenic and thermogenic methane sources. It can provide valuable information on methane origins and pathways.
9. Continuous Emission Monitoring Systems (CEMS)	It is installed at industrial facilities to continuously monitor methane emissions. It utilises sensors, analysers, and data acquisition systems to provide real-time measurements and emissions data.
10. Flux Towers	Tall towers equipped with sensors measure methane fluxes between the Earth's surface and the atmosphere over extended periods, providing continuous data on methane exchange dynamics.
11. Surface Seep Detection	Detects methane seeps on the Earth's surface using infrared cameras or other imaging techniques. It helps identify methane emission hotspots, such as natural gas leaks or geological seeps.
12. Passive Sampling	It involves deploying passive samplers, such as sorbent tubes or diffusive samplers, to collect methane over a specific period. Samples are then analysed to determine average concentrations.
13. Microbial DNA Analysis	Analyses microbial communities in soil or water samples to identify methane-producing microorganisms. Provides insights into the microbial processes driving methane emissions in natural environments.
14. Mobile Gas Chromatography	Utilises portable gas chromatographs to measure methane concentrations in ambient air or industrial emissions. It provides high-resolution data and can be deployed for field measurements.

Table 4.1 contd. ...

...Table 4.1 contd.

Technique	Description
15. Laser Absorption Spectroscopy	Employs lasers to measure methane concentrations based on the absorption of specific wavelengths of light by methane molecules. Offers high sensitivity and precision for atmospheric measurements.
16. Satellite Imagery	It uses satellite imagery to detect methane plumes and hotspots from space. This provides broader spatial coverage for identifying methane sources and monitoring large-scale emission trends.
17. Open-Path FTIR Spectroscopy	Measures methane concentrations by analysing the absorption of infrared radiation along open paths between a source and a detector. It is suitable for long-term monitoring and remote sensing applications.
18. Acoustic Monitoring	Detects methane emissions by capturing sound waves generated by escaping gas. Utilises specialised microphones or acoustic sensors to identify leaks or emission sources in industrial settings.
19. Radiocarbon Dating	It determines the age of methane emissions by measuring the decay of radiocarbon isotopes (carbon-14). This method is useful for distinguishing between old and recent methane emissions in atmospheric samples.
20. Methane Sniffers	Handheld devices equipped with methane sensors detect leaks or fugitive emissions in industrial facilities, pipelines, or natural gas infrastructure. They offer portability and rapid detection.
21. LIDAR (Light Detection and Ranging)	It uses laser pulses to measure methane concentrations in the atmosphere. It offers high spatial resolution and can penetrate vegetation, making it suitable for mapping methane emissions in diverse environments.
22. Portable Mass Spectrometry	Employs portable mass spectrometers to measure methane concentrations in ambient air or gas samples. Provides real-time data and high sensitivity for detecting low concentrations of methane.
23. Photoacoustic Spectroscopy	Measures methane concentrations based on the absorption of modulated light by methane molecules. It offers high sensitivity and selectivity for detecting methane in various environmental samples.
24. Wireless Sensor Networks	It consists of distributed sensor nodes deployed across a geographic area to monitor methane concentrations continuously. Data from individual nodes are transmitted wirelessly to a central database.
25. Cavity Ring-Down Spectroscopy	Measures methane concentrations based on light decay within a high-finesse optical cavity. It provides high sensitivity and precision for detecting low concentrations of methane in ambient air.
26. Differential Absorption LIDAR	It utilises differential absorption LIDAR techniques to measure methane concentrations in the atmosphere. It offers high sensitivity and can distinguish methane from other gases in complex environments.
27. Ultra-Portable Greenhouse Gas Analysers	Handheld analysers can measure methane concentrations accurately and precisely in various field settings. They are ideal for rapid assessments of methane emissions and source identification.
28. Smartphone Apps	Mobile applications equipped with methane sensors for citizen science initiatives or personal exposure monitoring. Provide user-friendly interfaces for data collection and sharing.

Table 4.1 contd. ...

...Table 4.1 contd.

Technique	Description
29. Bubble Strip Sampling	It utilises bubble strips or bubble traps to collect methane bubbles rising from aquatic environments. Samples are then analysed to determine methane concentrations and emission rates from water bodies.
30. Hyperspectral Imaging	It utilises hyperspectral cameras to capture images across multiple spectral bands, allowing for the detection and mapping of methane plumes in the atmosphere or on the Earth's surface.
31. Nanotechnology-based Sensors	It utilises nanomaterials and nanostructures to develop susceptible and selective methane sensors. These sensors offer miniaturisation, low power consumption, and rapid response for monitoring methane emissions.

4.6 Mitigation Strategies

Biogenic methane emissions from natural sources such as wetlands and agricultural activities significantly contribute to atmospheric methane concentrations. Implementing strategies to mitigate biogenic methane emissions is paramount to effectively diminishing their impact on climate change and achieving global emission reduction goals. This section explores different strategies to reduce biogenic methane emissions, focusing on approaches to wetland management and agricultural practices. Wetlands, including marshes, swamps, and peatlands, are significant in generating methane emissions, accounting for a proportion ranging from 20 to 40% of the total global methane emissions. Microbial mechanisms, namely acetoclastic and hydrogenotrophic methanogenesis, are crucial in facilitating methane generation within wetlands. The processes above are driven by the lack of oxygen and the breakdown of organic substances. Wetland management involves a comprehensive approach encompassing restoration, conservation, and mitigation strategies to reduce methane emissions. These measures are implemented to mitigate methanogenesis and enhance methane oxidation processes. Restoring and conserving natural wetland ecosystems can help reduce methane emissions by maintaining hydrological conditions, protecting vegetation cover, and promoting methane oxidation processes. Preserving wetlands is paramount in ensuring the long-term sustainability of wetland habitats and mitigating carbon and methane emissions into the atmosphere. This is achieved by preventing drainage, conversion, and disturbance activities. Restored wetlands can serve as carbon sinks by efficiently capturing atmospheric carbon dioxide through vegetation growth and organic matter accumulation. This process serves the dual purpose of mitigating methane emissions and bolstering climate resilience. The management of water levels and hydrological regimes in wetlands can impact methane emissions through the regulation of anaerobic conditions and methanogenic activity. The maintenance of aerobic conditions within wetland soils can be achieved by implementing controlled flooding, manipulating the water table, and using hydraulic engineering techniques. These techniques hinder methane generation and promote methane processing through oxidation by aerobic methanotrophic bacteria. Moreover, integrating artificial wetlands that include carefully planned components like baffles, channels, and sediment traps can enhance the ability to retain water and reduce methane emissions by facilitating better sedimentation and nutrient retention. The observation of methane emissions in wetlands can be influenced by managing vegetation composition and structure, affecting organic matter inputs, root exudation, and microbial communities. Encouraging the growth of non-methanogenic plant species, such as sedges, grasses, and shrubs, can reduce methane production by limiting substrate availability and promoting aerobic decomposition processes (Akam et al. 2023, García-Luna and Ortiz 2024, Hu et al. 2024, Lin et al. 2024, Ma et al. 2024, Minardi et al. 2023, Sayyah et al. 2022, Tan et al. 2024, C. Wang et al. 2024, X. Wang et al. 2024, J. Wang et al. 2024, Zhang et al. 2023, M. Zhang et al. 2024). The potential enhancement of methane absorption and conversion in wetland soils can be achieved by implementing vegetation species that demonstrate elevated methane oxidation rates, such as emergent aquatic plants and marsh grasses. Consequently, this can significantly mitigate methane emissions and facilitate carbon sequestration. Agricultural practices, encompassing livestock farming, rice cultivation, and manure management, significantly contribute to the release of biogenic methane, constituting approximately 40 to 45% of global anthropogenic methane emissions. Sustainable agricultural practices and management strategies can reduce methane emissions from agricultural sources, improving productivity, efficiency, and resilience. Ruminant animals, such as cattle, sheep, and goats, significantly influence methane emissions in livestock production. The main contributing factor to this phenomenon is enteric fermentation, which is the digestive process of ruminants and produces methane as a byproduct. To decrease methane emissions, one can include methane inhibitors, such as feed additives like 3-nitrooxypropanol, in the diet. These inhibitors suppress methanogenesis in the rumen. Further strategies for mitigating methane emissions from livestock include enhancing feed quality, optimising feeding practices, and improving animal health and productivity. Rice cultivation in flooded paddies is a significant source of methane emissions, accounting for approximately 10% of worldwide anthropogenic methane emissions. The potential mitigation of methane emissions from rice paddies can be achieved by implementing water-conservation strategies, such as alternate wetting and drying (AWD) irrigation and mid-season drainage. These strategies aim to reduce water volume and improve soil oxygenation. AWD irrigation involves periodically drying the soil between irrigation cycles, which helps reduce methane emissions by limiting the breakdown of organic substances without oxygen. In addition, the incorporation of aerobic rice cultivars and the advocacy for integrated rice-fish farming systems possess the capacity to enhance methane oxidation mechanisms and alleviate methane emissions linked to rice farming. Managing livestock manure offers a significant opportunity to mitigate methane emissions in the agricultural industry. In storage facilities such as lagoons, pits, and piles, methane is generated via the anaerobic decomposition of manure. Methane has the potential to be captured and utilised for energy generation through the implementation of anaerobic digestion or biogas recovery systems. Manure management practices such as covered storage, anaerobic digestion, and composting can effectively reduce methane emissions, produce renewable energy, improve nutrient management, and minimise the presence of odours and pathogens

(García-Luna and Ortiz 2024, Ghannadi and Dincer 2024, Ji and Shen 2024, Ma et al. 2024, Pintér 2024, Sayyah et al. 2022, Tan et al. 2024, Vargas-Rodríguez et al. 2024, C. Wang et al. 2024, Xu et al. 2024, Y. Zhang et al. 2024, Zhang et al. 2023, M. Zhang et al. 2024). The potential for enhancing soil health, fertility, and carbon sequestration from agricultural soils, as well as mitigating methane emissions, exists through the implementation of cover cropping and management practices. Cover crops such as legumes, grasses, and brassicas can suppress methane production by promoting aerobic soil conditions, enhancing organic matter decomposition, and stimulating microbial activity. Conservation tillage, crop rotation, and organic amendments can improve soil structure, water retention, and nutrient cycling and increase agricultural productivity and sustainability. These practices have the potential to mitigate methane emissions significantly. Effectively mitigating biogenic methane emissions from wetlands and agricultural activities requires implementing comprehensive strategies that address the multiple factors contributing to methane production and optimising the mechanisms involved in methane oxidation. Adopting wetland management strategies, including restoration, hydrological management, and vegetation control, proves advantageous in addressing methane emissions from natural wetland ecosystems. Similarly, adopting sustainable agricultural practices, such as livestock management, rice cultivation techniques, and manure management policies, can effectively reduce methane emissions resulting from agricultural operations while simultaneously promoting productivity, resilience, and sustainability. Through the integration of methane mitigation initiatives within broader conservation and agricultural development strategies, society has the potential to reduce its ecological footprint, mitigate the effects of climate change, and foster a more resilient and sustainable future. The levels of methane in the atmosphere and the occurrence of climate change are significantly influenced by anthropogenic sources of methane emissions, including waste management practices, the extraction and combustion of fossil fuels, and industrial processes (Akimbekov et al. 2024, Batgi and Dincer 2024, Cannone et al. 2024, García-Casado et al. 2024, Hadian et al. 2023, Huang et al. 2023, Kadam and Yadav 2024, Miao et al. 2022, Omidkar et al. 2023, Peng et al. 2024, Peng 2023, Prabowo et al. 2024, Tampio et al. 2024, Thalasso et al. 2022, Tian et al. 2024, Zhang et al. 2022, Zhou et al. 2023). The effective implementation of control measures aimed at mitigating methane emissions resulting from human activities is paramount in attaining climate stabilisation goals and mitigating the impacts of global warming. This section explores different strategies for controlling human-caused methane emissions, focusing on improving waste management techniques and implementing methods for capturing and utilising methane. Waste management practices, including landfill management, wastewater treatment, and organic waste recycling, significantly influence human-caused methane emissions. The effective reduction of methane emissions, facilitation of resource recovery, and mitigation of adverse environmental effects can be achieved by implementing enhanced waste management strategies. Landfills play a substantial role in the global release of methane, accounting for approximately 20 to 25% of all methane emissions attributed to human activities. Microbial activity during the anaerobic decomposition of organic waste deposited in landfills leads to the generation of methane as a byproduct. The effective capture and mitigation of methane emissions into the atmosphere can be achieved by implementing landfill gas management systems, such as landfill gas collection and flaring or utilisation. Gas collection systems in landfills typically consist of vertical or horizontal gas wells strategically placed within the cells of the landfill. The wells above are responsible for the extraction of gas that is rich in methane. This gas is then subjected to various processes and treatments, including flare-off or utilisation for energy generation through combustion or conversion into biogas. Wastewater treatment plants (WWTPs) significantly generate methane emissions from human activities, primarily attributable to the absence of oxygen during the natural decomposition of organic substances in sewage sludge. This biological process, known as anaerobic digestion, enables the decomposition of organic matter in the absence of oxygen. Methane is generated as a byproduct of this process, and it is typically either emitted into the atmosphere or flared along with the treatment. The combination of anaerobic digestion systems and biogas recovery and utilisation enables the efficient capture of methane emissions from sewage sludge and their conversion into a sustainable energy source. Using biogas produced through anaerobic digestion offers potential for on-site energy generation, heating, and transportation fuel, thereby reducing reliance on fossil fuels and mitigating the negative impacts of greenhouse gas emissions. Recycling organic waste, which includes composting, anaerobic digestion, and biomethanization, offers opportunities to divert organic waste from landfills and reduce methane emissions. Additionally, it can produce valuable soil amendments or renewable energy (García-Luna and Ortiz 2024, Ghannadi and Dincer 2024, He et al. 2023, Ma et al. 2024, Tan et al. 2024, Vargas-Rodríguez et al. 2024, C. Wang et al. 2024, Xu et al. 2024, Zhang et al. 2023, M. Zhang et al. 2024, Y. Zhang et al. 2024). Composting involves the deliberate and regulated aerobic decomposition of organic matter, leading to compost generation. This compost, in turn, functions as a nutrient-rich soil conditioner. The anaerobic digestion process involves using microorganisms to break down organic matter in an oxygen-deprived environment, generating biogas predominantly composed of methane and carbon dioxide. Biomethanization is a multifaceted process combining anaerobic digestion with mechanical separation, thermal hydrolysis, or microbial enhancement methods. These techniques are employed to improve the efficiency and quality of biogas production. The recycling of organic waste is of utmost importance in the ongoing efforts to reduce methane emissions, conserve landfill space, and promote the principles of a circular economy. This is achieved by redirecting organic waste away from landfills and enabling the recovery of valuable resources through composting and anaerobic digestion processes. Implementing methane capture and utilisation technologies offers promising opportunities to mitigate methane emissions from various human activities while harnessing methane as a valuable energy resource. The application of methane emissions in energy or product production has the potential to mitigate greenhouse gas emissions effectively, reduce reliance on fossil fuels, and promote sustainable development. Landfill gas recovery systems capture methane emissions from landfills, which convert them into renewable energy through either combustion or utilisation processes. To be used as a fuel source for engines, turbines or boilers, landfill gas, which contains a significant amount of methane, is typically processed to remove impurities such as hydrogen sulphide, moisture, and siloxanes. Projects focused on using landfill gas possess the capacity to generate electricity, heat, or steam, which can be utilised on-site or exported to the power grid. This has the potential to serve as a viable alternative to energy sources obtained from fossil fuels, thereby making a significant contribution towards mitigating greenhouse gas emissions. In addition, landfill gas recovery initiatives can generate revenue by selling Renewable Energy Credits (RECs) or carbon offsets, providing financial incentives for reducing methane emissions and recovering valuable resources (Akimbekov et al. 2024, Batgi and Dincer 2024, Cannone et al. 2024, García-Casado et al. 2024, Huang et al. 2023, Peng 2023, Peng et al. 2024, Prabowo et al. 2024, Tampio et al. 2024, Thalasso et al. 2022, Tian et al. 2024, Zhang et al. 2022, Zhou et al. 2023). The potential for harnessing and transforming methane emissions into sustainable energy or products is promising through the production of biogas derived from diverse organic waste sources, including agricultural residues, food waste, and wastewater. Anaerobic digestion systems are employed to decompose organic substances under controlled conditions, leading to the production of biogas, which is a mixture of methane and carbon dioxide, as well as digestate, a nutrient-rich byproduct. Biogas can be converted into biomethane using purification techniques such as gas scrubbing, Pressure Swing Adsorption (PSA), or membrane separation. Consequently, a highly pure methane product is obtained, ideal for injecting into natural gas pipelines or using as a fuel for transportation. Using biomethane as an alternative to fossil natural gas in the domains of heating, electricity generation, and vehicle fleets can effectively reduce greenhouse gas emissions and promote the complete elimination of carbon emissions within the energy sector. Methane hydrates are ice-like structures of methane molecules trapped in water crystals. They are large storage areas for methane kept underground or in permafrost regions. The stability and environmental hazards linked to methane hydrates pose substantial obstacles to energy extraction. Nevertheless, ongoing research efforts are primarily directed towards advancing technologies that efficiently extract and harness methane hydrates. Researchers have conducted recent studies to explore different methodologies, such as depressurisation, thermal stimulation, and carbon dioxide injection, as potential strategies for extracting methane from hydrate deposits. The objective of these investigations is to mitigate environmental impacts and minimise the occurrence of methane leakage (García-Luna and Ortiz 2024, He et al. 2023, Ma et al. 2024, Pintér 2024, Tan et al. 2024, C. Wang et al. 2024, L. Wang et al. 2024, Xu et al. 2024, M. Zhang et al. 2024, Zhang et al. 2023). The potential solution for bridging the gap between fossil fuels and renewable energy sources lies in extracting and utilising methane from hydrate deposits. This approach has the potential to effectively reduce greenhouse gas emissions and respond to the increasing energy demand. Direct Air Capture (DAC) technologies offer significant potential for directly capturing atmospheric methane emissions, facilitating their conversion into valuable products or storage. To capture methane molecules from the surrounding atmosphere and concentrate them for subsequent processing and deployment, direct air capture (DAC) systems utilise sorbents, membranes, or chemical reactions. Using captured methane as a primary resource for diverse industrial operations, fuel production, chemical replication, or subterranean sequestration through Carbon Capture and Storage (CCS) methodologies represent a feasible alternative. Despite being in its early stages of development and facing numerous technical and economic challenges, the direct air capture (DAC) of methane shows promise as a feasible approach to mitigate methane emissions and address the problem of climate change. Implementing advanced waste management practices and using methane capture and utilisation technologies offer promising opportunities for reducing human-caused methane emissions, promoting the recovery of resources, and strengthening energy security. Implementing strategies such as landfill gas recovery, biogas production, and methane hydrate recovery is recommended to mitigate methane emissions from diverse human sources and harness methane as a valuable energy resource. Society can integrate methane mitigation measures into waste management systems, energy infrastructure, and climate policy frameworks to address climate change challenges and advance towards a more sustainable and resilient future. Effectively mitigating methane emissions necessitates the implementation of coordinated policy interventions and international cooperation to foster innovation and attain climate stabilisation objectives. This section examines crucial policy strategies, global accords, and cooperative endeavours to diminish methane emissions and alleviate climate change. The development and implementation of policies aimed at reducing methane emissions in various sectors, such as energy, agriculture, waste management, and industry, are primarily influenced by national governments. Policies aimed at reducing methane emissions encompass a range of measures, such as regulatory actions, economic incentives, technological incentives, and public awareness campaigns (Batgi and Dincer 2024, Cannone et al. 2024, García-Casado et al. 2024, Huang et al. 2023, Kadam and Yadav 2024, Miao et al. 2022, Omidkar et al. 2023, Peng 2023, Peng et al. 2024). Legal requirements for reducing methane emissions in specific sectors or activities are established through regulatory measures, including emission standards, performance requirements, and pollution controls. Methane-intensive industries, such as oil and gas production, coal mining, and wastewater treatment, may be subject to government mandates requiring methane capture technologies, emission reduction targets, or the Best Available Techniques (BAT). Regulatory frameworks may encompass emissions reporting and verification provisions, which monitor adherence to regulations and evaluate advancements made towards achieving emission reduction objectives. Economic incentives, such as implementing carbon pricing mechanisms, provision of subsidies, tax credits, and allocation of grants, serve as financial motivators for engaging in methane mitigation activities and making investments. Carbon pricing mechanisms, such as carbon levies or emissions trading systems, incorporate the expenses associated with greenhouse gas emissions and establish financial motivations for reducing emissions. Providing subsidies and tax credits for renewable energy, energy efficiency, and methane capture initiatives can incentivise investment in environmentally friendly technologies and expedite shifting towards low-carbon energy alternatives. Governments have the potential to offer grants or financial assistance for Research and Development (R&D) endeavours, demonstration projects, and capacity-building initiatives that are specifically targeted at mitigating methane emissions. Technology incentives, such as financial support for research, grants for innovation, and programmes for transferring technology, facilitate the development and implementation of methane mitigation technologies and practices. Governments can allocate financial resources towards research institutions, universities, and private sector collaborators to foster the development of novel technologies for capturing, utilising, and monitoring methane (García-Luna and Ortiz 2024, Ghannadi and Dincer 2024, Ma et al. 2024, Pintér 2024, Sayyah et al. 2022, Tan et al. 2024, Vargas-Rodríguez et al. 2024, C. Wang et al. 2024, Xu et al. 2024, Y. Zhang et al. 2024, Zhang et al. 2023, M. Zhang et al. 2024). Technology transfer programmes play a crucial role in enabling the adoption of optimal practices and technologies in developing nations using capacity-building, training, and initiatives focused on technology transfer. The acceleration of innovation and technological advancements in methane mitigation can be facilitated through collaborative research and development partnerships involving government agencies, academia, and industry stakeholders. Public awareness campaigns, educational initiatives, and outreach endeavours enhance the understanding of methane emissions, climate change consequences, and the significance of methane mitigation among policymakers, businesses, and the wider public. Governments have the potential to initiate public awareness campaigns via various channels such as media outlets, educational institutions, and community organisations. These campaigns disseminate information, encourage behavioural modifications, and garner support for implementing methane mitigation measures. Public consultations, stakeholder dialogues, and participatory decision-making processes can effectively involve stakeholders, promote consensus, encourage collaboration, and improve the efficacy of methane mitigation policies. Global climate action and the mitigation of methane emissions heavily rely on international collaboration and multilateral agreements. International agreements serve as structures for fostering collaboration, facilitating knowledge exchange, and enhancing capacity-building endeavours to tackle shared obstacles and attain shared objectives for reducing emissions. The UNFCCC, established in 1992, is the comprehensive global climate action and collaboration framework. The objective of the UNFCCC is to achieve a state of equilibrium in greenhouse gas levels in the atmosphere, ensuring that human activities do not significantly disrupt the climate system (Akam et al. 2023, García-Luna and Ortiz 2024, Ghannadi and Dincer 2024, He et al. 2023, Lin et al. 2024, Litheko et al. 2023, Minardi et al. 2023, Pintér 2024, Sabrekov et al. 2024, X. Wang et al. 2024, M. Zhang et al. 2024). The parties to the United Nations Framework Convention on Climate Change (UNFCCC) pledge to decrease greenhouse gas emissions, strengthen climate resilience, and advance sustainable development using Nationally Determined Contributions (NDCs). These NDCs delineate each country's climate mitigation and adaptation objectives. Methane mitigation plays a crucial role in attaining the goals set forth by the United Nations Framework Convention on Climate Change (UNFCCC). This objective is tackled through various mechanisms, such as emissions reporting, technology transfer, capacity-building, and financial assistance. The Paris Agreement, ratified in 2015 within the framework of the United Nations Framework Convention on Climate

Change (UNFCCC), establishes ambitious objectives to restrict the rise in global temperatures to less than 2 degrees Celsius above pre-industrial levels while also striving to limit the temperature increase to 1.5 degrees Celsius. According to the Paris Agreement, a comprehensive structure facilitates global collaboration in climate mitigation, adaptation, finance, and technology transfer. Signatories to the Paris Agreement provide Nationally Determined Contributions (NDCs) that detail their obligations to decrease greenhouse gas emissions, specifically methane, and strengthen climate resilience. The Agreement additionally incorporates provisions for transparency, accountability, and global stocktaking mechanisms to evaluate advancements made towards emission reduction objectives and enhance ambition in the long run. The Kyoto Protocol was ratified in 1997 as an appendix to the United Nations Framework Convention on Climate Change (UNFCCC). It introduced obligatory targets for reducing emissions for developed nations. It established adaptable mechanisms for trading emissions, implementing joint initiatives, and promoting clean development projects. The Kyoto Protocol primarily focused on carbon dioxide emissions. Still, it also tackled other greenhouse gases, such as methane, by committing to reduce emissions and providing incentives for projects to mitigate methane. The Kyoto Protocol's Clean Development Mechanism (CDM) facilitated methane mitigation initiatives in developing nations by providing monetary incentives to reduce emissions. Collaborative endeavours and alliances facilitate the convergence of governmental bodies, enterprises, Non-Governmental Entities (NGOs), research establishments, and additional relevant parties to tackle methane emissions through collective action, exchanging knowledge, and enhancing capabilities. The Global Methane Initiative (GMI), initiated in 2004, is a collaborative effort among governments, businesses, non-governmental organisations (NGOs), and various stakeholders. Its primary objective is to decrease methane emissions and promote the development of methane mitigation technologies and practices. GMI directs its efforts towards addressing significant sources of methane emissions, such as agriculture, coal mines, municipal solid waste, and oil and gas operations, using collaborative projects, technical assistance, and capacity-building activities. GMI facilitates sharing information, technology transfer, and policy assistance provision to expedite efforts to reduce methane emissions and attain emission reduction goals. Established in 2012, the Climate and Clean Air Coalition (CCAC) is a collaborative initiative comprising governmental bodies, corporate entities, and global institutions. Its primary objective is to mitigate the impact of Short-Lived Climate Pollutants (SLCPs), such as methane, black carbon, and hydrofluorocarbons (HFCs). The primary objective of CCAC is to effectively execute tangible strategies to mitigate SLCP emissions originating from key sectors, including agriculture, waste management, transportation, and energy production. The CCAC helps countries implement methane mitigation measures and transition to low-emission pathways by facilitating policy dialogue, capacity-building, and technical assistance. The international initiative, the Methane to Markets Partnership (M2M), initiated in 2004, aims to mitigate methane emissions and facilitate the development of projects focused on the recovery and utilisation of methane (Akimbekov et al. 2024, Batgi and Dincer 2024, Cannone et al. 2024, Huang et al. 2023, Omidkar et al. 2023, Peng

Table 4.2. Different types of mitigation strategies for methane removal from the environment (Source: Eisavi et al. 2024, García-Luna and Ortiz 2024, Hu et al. 2024, Litheko et al. 2023, Ma et al. 2024, Mao et al. 2024, Minardi et al. 2023, Sun et al. 2023).

Mitigation Strategy	**Description**
1. Anaerobic Digestion	Converts organic waste, such as agricultural residues, manure, and food waste, into biogas through microbial decomposition in anaerobic conditions. Biogas produced can be used as a renewable energy source, reducing methane emissions from waste decomposition.
2. Landfill Gas Capture and Utilisation	Collects methane emitted from landfills using gas collection systems. Captured methane can be utilised for energy generation through combustion in engines or turbines, reducing methane emissions and offsetting the use of fossil fuels.
3. Manure Management Practices	Implements practices such as covered lagoons, anaerobic digestion, and composting to manage livestock manure effectively. These practices help reduce methane emissions from manure decomposition and can also produce biogas for energy generation.
4. Methane Oxidation Catalysts	It utilises catalysts to promote methane oxidation to carbon dioxide and water in industrial processes or natural environments. Methane oxidation reduces methane emissions and converts methane into less potent greenhouse gases.
5. Rice Cultivation Techniques	Alternate wetting and drying (AWD) irrigation and mid-season drainage in rice paddies are adopted to reduce methane emissions. These techniques modify soil conditions to minimise anaerobic decomposition and methane production.
6. Methane Capture from Livestock	Installs methane capture systems in livestock facilities to collect methane emitted from enteric fermentation and manure decomposition. Captured methane can be used for energy production, reducing emissions from livestock operations.
7. Improved Waste Management	Implements advanced waste management practices, including recycling, composting, and waste-to-energy technologies, to reduce organic waste sent to landfills. Reducing organic waste decreases methane emissions from anaerobic decomposition in landfills.
8. Methane Hydrate Extraction	Extracts methane hydrates from marine sediments or permafrost regions to prevent their release into the atmosphere. Controlled extraction can mitigate the risk of sudden methane releases, reducing the potential for exacerbating climate change.
9. Natural Wetland Restoration	Restores degraded or drained wetland ecosystems to their natural state to enhance methane uptake and storage. Restored wetlands act as carbon sinks, sequestering methane and reducing atmospheric methane concentrations.
10. Improved Enteric Fermentation Management	Implements dietary adjustments, feed additives, and improved animal husbandry practices to reduce methane emissions from enteric fermentation in ruminant animals. These management strategies aim to optimise digestion and reduce methane production.

Table 4.2 contd. ...

...Table 4.2 contd.

Mitigation Strategy	Description
11. Controlled Biomass Burning	Implements controlled burning practices for agricultural residues, forest debris, and other biomass sources to minimise methane emissions. Controlled burns can reduce methane release during decomposition and promote more efficient combustion.
12. Methane Capture from Wastewater Treatment	Captures methane produced during wastewater treatment processes, such as anaerobic digestion of sewage sludge. Captured methane can be used for energy generation or converted into less potent greenhouse gases, reducing emissions from treatment facilities.
13. Improved Fossil Fuel Extraction and Handling	Enhances leak detection and repair (LDAR) programs, improves equipment maintenance practices, and reduces venting and flaring during fossil fuel extraction and processing operations. These measures minimise methane emissions from fossil fuel operations.
14. Methane Mitigation in Oil and Gas Production	Implements technologies such as green completions, Vapour Recovery Units (VRUs), and pneumatic device replacements to capture and reduce methane emissions from oil and gas production operations. Methane mitigation measures help minimise fugitive emissions and leakage.
15. Enhanced Agricultural Practices	Adopts sustainable agricultural practices, including precision farming, agroforestry, and cover cropping, to improve soil health and reduce methane emissions from agricultural activities. These practices enhance carbon sequestration and minimise methane release.
16. Methane Capture from Coal Mines	Installs methane capture and utilisation systems in coal mines to capture methane released during mining operations. Captured methane can be used for energy generation or converted into other products, reducing methane emissions from coal mining activities.
17. Improved Industrial Process Efficiency	Enhancing energy efficiency, implementing process optimisation, and adopting cleaner production technologies reduce methane emissions from industrial processes. Improving process efficiency minimises energy consumption and associated methane emissions.
18. Methane Recovery from Anaerobic Ponds	Utilises methane recovery systems in anaerobic ponds, such as covered lagoons or floating covers, to capture methane emitted from wastewater treatment facilities. Captured methane can be used for energy production or other applications.
19. Urban Methane Leak Detection and Repair	Implements methane leak detection and repair programs in urban areas to identify and address methane leaks from natural gas distribution systems, pipelines, and infrastructure. Prompt leak repair helps minimise methane emissions and ensures public safety.
20. Enhanced Methane Destruction Technologies	Develops and deploys advanced methane destruction technologies, such as thermal oxidation, plasma torches, and catalytic combustion, to efficiently convert methane into less potent greenhouse gases or harmless byproducts. Enhanced destruction reduces methane emissions.
21. Methane Mitigation in Land Use Changes	Minimising deforestation prevents wetland drainage and promotes reforestation and afforestation to preserve natural carbon sinks and reduce methane emissions associated with land use changes. Conservation efforts help mitigate methane release from altered ecosystems.

Table 4.2 contd. ...

...Table 4.2 contd.

Mitigation Strategy	Description
22. Carbon Capture and Storage (CCS)	Captures methane emissions from industrial sources, power plants, or natural gas processing facilities and stores the captured carbon underground to prevent its release into the atmosphere. CCS technologies help reduce methane emissions and mitigate climate change.
23. Methane Mitigation in Coal Bed Methane Production	Implements best management practices, such as degasification systems and methane drainage, to capture and reduce methane emissions from coal bed methane extraction operations. Methane mitigation measures minimise emissions during coal bed methane production.
24. Renewable Energy Integration	Promotes the integration of renewable energy sources, such as solar, wind, and hydroelectric power, to displace fossil fuel-based energy generation. Shifting to renewable energy reduces methane emissions associated with fossil fuel combustion.
25. Agricultural Residue Management	Utilises agricultural residues for energy production through biomass combustion, anaerobic digestion, or biofuel production. Effective residue management reduces methane emissions from decomposition and provides renewable energy sources.
26. Wetland Conservation and Restoration	Preserve existing wetlands, restore degraded wetland ecosystems, and create new wetland habitats to enhance methane uptake and storage. Wetland conservation and restoration efforts help sequester methane and mitigate atmospheric methane concentrations.
27. Methane Capture from Digesters	Captures methane generated from anaerobic digesters used for organic waste treatment, such as sewage sludge digestion or industrial effluent treatment. Captured methane can be utilised for energy production or converted into other valuable products.
28. Methane Reduction in Cement Production	Adopts alternative raw materials, fuel switching, and energy efficiency measures to reduce methane emissions from cement production processes. Improved production practices help minimise methane release and lower the carbon footprint of cement manufacturing.
29. Methane Mitigation in Peatland Restoration	Implements peatland restoration and conservation projects to prevent peat degradation and minimise methane emissions from drained peatlands. Restoration efforts aim to preserve peatland carbon stores and reduce methane release from degraded ecosystems.
30. Enhanced Methane Mitigation Policies	Develops and enforces policies, regulations, and incentives to promote methane mitigation measures across various sectors. Enhanced policy frameworks encourage industry adoption of mitigation technologies and support emissions reduction goals.
31. Public Awareness and Education	Raises public awareness about methane emissions, climate change impacts, and mitigation strategies through education campaigns, outreach programs, and community engagement. Increased awareness fosters support for methane reduction initiatives and encourages behavioural changes.

et al. 2024, Prabowo et al. 2024, Thalasso et al. 2022, Tian et al. 2024, Zhang et al. 2022, Zhou et al. 2023). Machine-to-Machine (M2M) technology facilitates collaboration among governmental bodies, enterprises, and technical specialists to identify potential methane capture and utilisation avenues in various sectors, including agriculture, coal mining, municipal solid waste management, and oil and gas extraction. M2M offers technical aid, financial assistance, and platforms for sharing knowledge to facilitate the global development and execution of methane mitigation projects. The International Energy Agency (IEA) assists with global initiatives to mitigate methane emissions from the energy industry by conducting research, conducting analysis, and offering policy suggestions. The IEA evaluates methane emissions from fossil fuel production, transmission, and consumption. It also guides the most effective methods, technologies, and policies for mitigating methane. The International Energy Agency (IEA) works with governments, industry stakeholders, and international organisations to raise awareness, enhance capabilities, and facilitate the exchange of knowledge regarding strategies to reduce methane emissions. To effectively address methane emissions, promote climate action, and attain global emission reduction goals, it is imperative to implement policy interventions and foster international cooperation. Methane mitigation measures are implemented based on national policy frameworks, while international agreements and collaborative initiatives facilitate cooperation, knowledge sharing, and capacity-building efforts across borders. Through the establishment of multilateral partnerships, governments, businesses, and stakeholders can collaboratively efficiently tackle methane emissions, foster sustainable development, and construct a future that is both resilient and characterised by reduced carbon emissions (Table 4.2).

4.7 Future Directions and Challenges

Amidst the pressing issue of climate change, methane mitigation is recognised as a crucial element in worldwide endeavours to decrease greenhouse gas emissions and restrict global warming. The future trajectory of methane mitigation necessitates a comprehensive strategy encompassing scientific inquiry, technological advancements, and socio-economic and political factors. This section analyses the primary areas of research focus, technological advancements, and the socio-economic and political obstacles that hinder the successful implementation of methane mitigation strategies. A comprehensive comprehension of the mechanisms that regulate methane generation, utilisation, and movement within the Earth's ecological systems is imperative to precisely measure methane emissions, evaluate their influence on the carbon cycle, and pinpoint potential avenues for mitigation. This knowledge is crucial for furthering our knowledge of biogeochemical processes that govern methane production, consumption, and transportation in diverse settings such as wetlands, soils, and oceans. Research should prioritise the investigation of microbial pathways, environmental factors, and feedback mechanisms that impact methane dynamics in terrestrial and aquatic environments. Additionally, efforts should be made to develop remote sensing technologies and monitoring techniques that can accurately and comprehensively measure methane emissions at regional and global levels. Incorporating satellite observations, aerial surveys, ground-based monitoring networks, and modelling simulations can augment our capacity to identify and ascribe methane origins, sinks, and patterns across various geographical areas and industries. Researchers are currently examining the comparative impacts of natural and anthropogenic origins on methane emissions. These sources encompass wetlands, agriculture, fossil fuel consumption, waste disposal, and industrial operations. Researchers should prioritise their efforts towards the identification of areas with high emissions, the quantification of factors contributing to emissions, the evaluation of the efficacy of mitigation strategies for various sources and environments of methane, and the examination of methane's involvement in climate feedback mechanisms and its interactions with other elements of the Earth's system, such as the carbon cycle, hydrological cycle, and atmospheric chemistry. The investigation of methane-climate feedback can enhance our comprehension of the consequences of methane emissions on climate sensitivity, extreme weather occurrences, and ecological dynamics. This research can contribute to developing climate models and policy choices and assess methane mitigation strategies' efficacy, feasibility, and additional advantages. These strategies encompass natural climate solutions, technological advancements, and policy interventions. Studying the potential for reducing methane emissions can provide valuable insights for decision-making, help determine the most important investment strategies, and direct the creation of comprehensive mitigation plans that tackle various environmental, social, and economic goals. Technological innovation is vital in creating affordable, expandable, and environmentally friendly solutions for reducing methane emissions in various industries and areas. To optimise the reduction of methane emissions and simultaneously generate renewable energy, heat, or value-added products, there is a need to enhance methane capture and utilisation technologies, including biogas recovery, anaerobic digestion, and biomethanization. The enhancement of efficiency, reliability, and economics in methane conversion processes can be achieved by investigating novel catalysts, membrane materials, and process optimisation techniques. Researchers are currently developing sensor technologies, remote sensing platforms, and monitoring systems to detect and quantify methane emissions from various sources and environments. Studying miniaturised sensors, drone-based monitoring, and satellite-based imaging can improve our capacity to detect areas with high emissions levels, monitor emissions patterns, and confirm real-time reductions in emissions. There is an ongoing investigation into various biological, chemical, and physical approaches aimed at augmenting the processes of methane oxidation and reduction in both natural and engineered systems. Methane mitigation in agricultural, industrial, and waste management settings can be facilitated by researching methane-consuming bacteria, biofiltration systems, and catalytic converters. These technologies can reduce emissions, improve air quality, and safeguard public health. Additionally, there is a need to explore carbon capture and storage (CCS) technologies for capturing methane emissions from fossil fuel production, processing, and utilisation facilities. Investigating carbon capture methodologies, including adsorption, absorption, and membrane separation, can potentially facilitate the economically viable capture of methane emissions. These emissions can subsequently be conveyed and sequestered beneath geological formations, resulting in a reduction of greenhouse gas emissions and the amelioration of climate change consequences. Integrating methane mitigation technologies with renewable energy systems, such as wind, solar, and hydroelectric power, yields synergistic effects and optimises environmental advantages. Investigating hybrid energy systems, energy storage solutions, and grid integration strategies can expedite the shift towards low-carbon energy sources, concurrently mitigating methane emissions from fossil fuel combustion and energy generation. Implementing methane mitigation strategies encounters socio-economic and political obstacles that necessitate attention to expedite advancements towards achieving emission reduction objectives. Potential avenues for tackling socio-economic and political obstacles encompass establishing a conducive environment for methane reduction efforts by aligning national policies, regulations, and incentives with climate mitigation goals. Facilitating methane mitigation measures and promoting sustainable development pathways can be achieved through enhancing policy coherence, coordination, and integration across various sectors, including energy, agriculture, transportation, and waste management. This study aims to explore using innovative financial mechanisms, including green bonds, carbon markets, and climate funds, to secure public and private sector financing for methane mitigation projects. To enhance private investment, facilitate capital flows, and expand methane mitigation initiatives in developing countries and emerging economies, it is imperative to implement dedicated funding mechanisms, risk-sharing arrangements, and blended finance models. The objective is to enhance the adoption of methane mitigation technologies in developing countries and regions with limited resources and expertise by facilitating technology transfer and capacity-building initiatives. Facilitating the spread of technology, sharing knowledge, and collaborating between different regions can help close the gaps in technology, enhance local capabilities, and expedite the implementation of cost-efficient solutions to mitigate methane emissions. Participating in decision-making processes, planning endeavours, and implementation activities about methane mitigation involves the active involvement of various stakeholders, such as governments, businesses, communities, and civil society organisations. The enhancement of ownership, transparency, and accountability in methane reduction initiatives can be achieved by cultivating multi-stakeholder partnerships, participatory approaches, and social inclusion. These strategies aim to ensure the representation and addressing of diverse perspectives and interests. Enhancing global cooperation and collaboration in methane mitigation comprises establishing bilateral agreements, multilateral initiatives, and global partnerships. Facilitating the exchange of information, providing technical aid, and offering support to develop skills can promote reciprocal knowledge exchange, spreading innovative ideas, and collaborative efforts to reduce methane emissions. This can contribute to achieving global climate goals and sustainable development objectives. The effective management of methane mitigation necessitates the establishment of interdisciplinary partnerships, developing inventive approaches, and implementing collective efforts on a global, national, and local scale. Society can expedite progress in reducing methane emissions, mitigating climate change impacts, and constructing a more sustainable and resilient future for the coming generations by prioritising research, investing in technological innovation, and overcoming socioeconomic and political obstacles.

4.8 Conclusion

In conclusion, methane production plays a significant role in the global carbon cycle, influencing atmospheric composition, climate dynamics, and ecosystem processes. As a potent greenhouse gas, methane contributes to radiative forcing and global warming, amplifying the impacts of anthropogenic activities on the Earth's climate system. Understanding the sources, sinks, and biogeochemical pathways of methane is essential for accurately quantifying methane emissions, assessing their impact on climate change, and identifying opportunities for mitigation. Natural sources such as wetlands, oceans, termites, and microbial processes contribute to biogenic methane emissions. In contrast, anthropogenic activities such as agriculture, fossil fuel extraction, waste management, and industrial processes contribute to anthropogenic methane emissions. Addressing methane emissions requires a comprehensive approach integrating scientific research, technological innovation, policy interventions, and international cooperation. Efforts to mitigate methane emissions must prioritise research on methane production, consumption, and feedback mechanisms within the Earth's ecosystems. The potential for technological innovations in methane capture, utilisation, detection, and reduction is promising, offering a beacon of hope in reducing emissions from diverse sources and sectors while promoting sustainable development and energy transition. Policy interventions, financial mechanisms, and stakeholder engagement are essential for creating an enabling environment for methane mitigation and scaling up mitigation efforts globally. International cooperation and collaboration are beneficial and crucial for effectively addressing methane emissions. Methane emissions transcend national borders and require coordinated action at the global scale. Multilateral agreements such as the Paris Agreement provide frameworks for countries to set emission reduction targets, implement mitigation measures, and enhance transparency and accountability in climate action, emphasising the need for global unity in this endeavour. In conclusion, addressing methane production and its role in the worldwide carbon cycle requires concerted efforts from governments, businesses, civil society, and the scientific community. By prioritising methane mitigation, investing in research and innovation, and fostering international cooperation, society can mitigate climate change impacts, protect ecosystems, and build a more sustainable and resilient future for the planet.

List of Abbreviation

1. CH_4 - Methane
2. CO_2 - Carbon dioxide
3. GHG - Greenhouse gas
4. GWP - Global warming potential
5. OH - Hydroxyl radicals

6. AWD - Alternate wetting and drying
7. LDAR - Leak detection and repair
8. VOCs - Volatile organic compounds
9. NOx - Nitrogen oxides
10. CO - Carbon monoxide
11. Aerenchyma
12. CO2e - Carbon dioxide equivalent
13. Gt - Gigaton
14. AWG - Anthropogenic warming potential
15. ICE - Intergovernmental Panel on Climate Change
16. Tg - Teragram
17. N_2O - Nitrous oxide
18. IPCC - Intergovernmental Panel on Climate Change
19. LULUCF - Land use, land-use change, and forestry
20. MW - Methane warming potential
21. CCS - Carbon capture and storage

References

Akam, Sajjad A., Elizabeth D. Swanner, Hongming Yao, Wei-Li Hong and Jörn Peckmann. (2023). Methane-derived authigenic carbonates—a case for a globally relevant marine carbonate factory. Earth-Science Reviews 243: 104487. doi: https://doi.org/10.1016/j.earscirev.2023.104487.

Akimbekov, Nuraly S., Ilya Digel, Kuanysh T. Tastambek, Marzhan Kozhahmetova, Dinara K. Sherelkhan and Zhandos Tauanov. (2024). Hydrogenotrophic methanogenesis in coal-bearing environments: methane production, carbon sequestration, and hydrogen availability. International Journal of Hydrogen Energy 52: 1264–77. doi: https://doi.org/10.1016/j.ijhydene.2023.09.223.

Batgi, Sibel Uygun and Ibrahim Dincer. (2024). A study on comparative environmental impact assessment of thermochemical cycles and steam methane reforming processes for hydrogen production processes. Computers & Chemical Engineering 180: 108514. doi: https://doi.org/10.1016/j.compchemeng.2023.108514.

Bidart, Christian, Martin Wichert, Gunther Kolb and Michael Held. (2022). Biogas catalytic methanation for biomethane production as fuel in freight transport—a carbon footprint assessment. Renewable and Sustainable Energy Reviews 168: 112802. doi: https://doi.org/10.1016/j.rser.2022.112802.

Cannone, Salvatore F., Muhammad Ishaq, Andrea Lanzini and Massimo Santarelli. (2024). Thermodynamic analysis of a synergistic integration of solid oxide fuel cell and solar-based chemical looping methane reforming unit for solar energy storage, power production, and carbon capture. Energy Conversion and Management 302: 118080. doi: https://doi.org/10.1016/j.enconman.2024.118080.

Eisavi, Beneta, Hossein Nami, Faramarz Ranjbar and Ali Sharifi. (2024). Economic assessment and optimization of low-carbon biomass-based power, methane, and methanol production. International Journal of Hydrogen Energy 52: 869–88. doi: https://doi.org/10.1016/j.ijhydene.2023.11.025.

García-Casado, Silvia, Raúl Muñoz and Raquel Lebrero. (2024). Enrichment of a mixed syngas-converting culture for volatile fatty acids and methane production. Bioresource Technology 130646. doi: https://doi.org/10.1016/j.biortech.2024.130646.

García-Luna, S. and Ortiz, C. (2024). Partial oxycombustion and amines-driven waste-to-methane for improved carbon capture and utilization (CCU). Energy Conversion and Management 302: 118125. doi: https://doi.org/10.1016/j.enconman.2024.118125.

Ghannadi, Mitra and Ibrahim Dincer. (2024). A new system for hydrogen and methane production with post-combustion amine-based carbon capturing option for better sustainability. Energy Conversion and Management 301: 118026. doi: https://doi.org/10.1016/j.enconman.2023.118026.

Hadian, Morteza, Kay Buist and Hans Kuipers. (2023). An overview of production of hydrogen and carbon nanomaterials via thermocatalytic decomposition of methane. Current Opinion in Chemical Engineering 42: 100968. doi: https://doi.org/10.1016/j.coche.2023.100968.

He, Yangdong, Bin Song, Xingsheng Jing, Yin Zhou, Honggang Chang, Wei Yang and Zeai Huang. (2023). Low-carbon hydrogen production via molten salt methane pyrolysis with chemical looping combustion: emission reduction potential and techno-economic assessment. Fuel Processing Technology 247: 107778. doi: https://doi.org/10.1016/j.fuproc.2023.107778.

Hu, Jia, Mathilde Bettembourg, Lihong Xue, Ronggui Hu, Anna Schnürer, Chuanxin Sun et al. (2024). A low-methane rice with high-yield potential realized via optimized carbon partitioning. Science of The Total Environment 920: 170980. doi: https://doi.org/10.1016/j.scitotenv.2024.170980.

Huang, Juan, Wurong Zhao, Jinwei Ju, Suifen Liu, Jinshao Ye and Yan Long. (2023). The existence of ferric hydroxide links the carbon and nitrogen cycles by promoting nitrite-coupled methane anaerobic oxidation. Water Research 243: 120192. doi: https://doi.org/10.1016/j.watres.2023.120192.

Ji, Jinqing and Laihong Shen. (2024). Enhanced co-production of high-quality syngas and highly-concentrated hydrogen via chemical looping steam methane reforming over ni-substituted La0.6Ce0.4MnO3 oxygen carriers. Fuel 368: 131588. doi: https://doi.org/10.1016/j.fuel.2024.131588.

Kadam, Ramdas S. and Ganapati D. Yadav. (2024). Life cycle analysis of ammonia and methane production using green hydrogen and carbon dioxide. Journal of Cleaner Production 449: 141620. doi: https://doi.org/10.1016/j.jclepro.2024.141620.

Keller, Jason K., Scott D. Bridgham, Kimberly K. Takagi, Cassandra A. Zalman, Jessica E. Rush, Crisand Anderson et al. (2023). Microbial organic matter reduction regulates methane and carbon dioxide production across an ombrotrophic-minerotrophic peatland gradient. Soil Biology and Biochemistry 182: 109045. doi: https://doi.org/10.1016/j.soilbio.2023.109045.

Lin, Chiao-Wen, Wei-Jen Lin, Chuan-Wen Ho, Yu-Chen Kao, Zhao-Jun Yong and Hsing-Juh Lin. (2024). Flushing emissions of methane and carbon dioxide from mangrove soils during tidal cycles. Science of The Total Environment 919: 170768. doi: https://doi.org/10.1016/j.scitotenv.2024.170768.

Litheko, Andrew, Bilainu Oboirien and Bilal Patel. (2023). Life cycle assessment of power-to-gas (PtG) technology – evaluation of system configurations of renewable hydrogen and methane production. Sustainable Energy Technologies and Assessments 60: 103527. doi: https://doi.org/10.1016/j.seta.2023.103527.

Ma, Rui, Ji Li, Tyagi, R. D. and Xiaolei Zhang. (2024). Carbon dioxide and methane as carbon source for the production of polyhydroxyalkanoates and concomitant carbon fixation. Bioresource Technology 391: 129977. doi: https://doi.org/10.1016/j.biortech.2023.129977.

Mao, Yufeng, Tong Lin, Hong Li, Ruixu He, Kailai Ye, Weiwei Yu and Qiang He. (2024). Aerobic methane production by phytoplankton as an important methane source of aquatic ecosystems: reconsidering the global methane budget. Science of The Total Environment 907: 167864. doi: https://doi.org/10.1016/j.scitotenv.2023.167864.

Miao, Xiaoming, Xiting Liu, Qing Li, Ang Li, Feng Cai, Fanxing Kong et al. (2022). Porewater geochemistry indicates methane seepage in the okinawa trough and its implications for the carbon cycle of the subtropical west pacific. Palaeogeography, Palaeoclimatology, Palaeoecology 607: 111266. doi: https://doi.org/10.1016/j.palaeo.2022.111266.

Minardi, Matteo, Paolo Marocco and Marta Gandiglio. (2023). Carbon recovery from biogas through upgrading and methanation: a techno-economic and environmental assessment. Journal of CO_2 Utilization 78: 102632. doi: https://doi.org/10.1016/j.jcou.2023.102632.

Mostafa, Abdelrahman, Irene Rapone, Aldo Bosetti, Matteo C. Romano, Alessandra Beretta and Gianpiero Groppi. (2023). Modelling of methane sorption enhanced reforming for blue hydrogen production in an adiabatic fixed bed reactor: unravelling the role of the reactor's thermal behavior. International Journal of Hydrogen Energy 48(68): 26475–91. doi: https://doi.org/10.1016/j.ijhydene.2023.03.357.

Omidkar, Ali, Hao Xu, Zhaofei Li, Kamran Haddadian and Hua Song. (2023). Techno-economic and life cycle assessment of renewable diesel production via methane-assisted catalytic waste cooking oil upgrading. Journal of Cleaner Production 414: 137512. doi: https://doi.org/10.1016/j.jclepro.2023.137512.

Oni, A. O., Giwa, T., Font-Palma, C. and Fadare, D. A. (2023). Comparative techno-economic and life cycle greenhouse gas assessment of ammonia production from thermal decomposition of methane and steam methane reforming technologies. International Journal of Greenhouse Gas Control 123: 103819. doi: https://doi.org/10.1016/j.ijggc.2022.103819.

Peng, Shushi. (2023). Challenges and opportunities in the global methane cycle. IScience 26(6): 106878. doi: https://doi.org/10.1016/j.isci.2023.106878.

Peng, Weijie, Jinrong Lu, Jiajie Kuang, Rong Tang, Fengyi Guan, Kunting Xie et al. (2024). Enhancement of hydrogenotrophic methanogenesis for methane production by nano zero-valent iron in soils. Environmental Research 247: 118232. doi: https://doi.org/10.1016/j.envres.2024.118232.

Pintér, Gábor. (2024). The development of global power-to-methane potentials between 2000 and 2020: a comparative overview of international projects. Applied Energy 353: 122094. doi: https://doi.org/10.1016/j.apenergy.2023.122094.

Prabowo, Justin, Leo Lai, Benjamin Chivers, Declan Burke, An Huy Dinh, Linlin Ye et al. (2024). Solid carbon co-products from hydrogen production by methane pyrolysis: current understandings and recent progress. Carbon 216: 118507. doi: https://doi.org/10.1016/j.carbon.2023.118507.

Qi, Meng, Yi Liu, Tianbiao He, Liang Yin, Chi-Min Shu and Il Moon. (2022). System perspective on cleaner technologies for renewable methane production and utilisation towards carbon neutrality: principles, techno-economics, and carbon footprints. Fuel 327: 125130. doi: https://doi.org/10.1016/j.fuel.2022.125130.

Sabrekov, Aleksandr F., Mikhail V. Semenov, Irina E. Terentieva, George S. Krasnov, Sergey L. Kharitonov, Mikhail V. Glagolev et al. (2024). Anaerobic methane oxidation is quantitatively important in deeper peat layers of boreal peatlands: evidence from anaerobic incubations, *in situ* stable isotopes depth profiles, and microbial communities. Science of The Total Environment 916: 170213. doi: https://doi.org/10.1016/j.scitotenv.2024.170213.

Sayyah, Ali, Mohammad Ahangari, Jafar Mostafaei, Seyed Reza Nabavi and Aligholi Niaei. (2023). Machine learning-based life cycle optimization for the carbon dioxide methanation process: achieving environmental and productivity efficiency. Journal of Cleaner Production 426: 139120. doi: https://doi.org/10.1016/j.jclepro.2023.139120.

Sayyah, Ali, Elham Mahmoudi, Samira Farhoudi, Gamze Behmenyar, Abdullah Zahid Turan, Seyed Reza Nabavi et al. (2022). Environmental assessment of carbon dioxide methanation process using mixed metal oxide and zeolite-supported catalysts by life cycle assessment methodology. Journal of Cleaner Production 362: 132529. doi: https://doi.org/10.1016/j.jclepro.2022.132529.

Su, Bosheng, Yilin Wang, Zhilong Xu, Wei Han, Hongguang Jin and Hongsheng Wang. (2022). Novel ways for hydrogen production based on methane steam and dry reforming integrated with carbon capture. Energy Conversion and Management 270: 116199. doi: https://doi.org/10.1016/j.enconman.2022.116199.

Sun, Jiaji, Eldon R. Rene, Yuhe He, Weifang Ma, Qian Hu and Bin Qiu. (2023). Carbon, iron, and polymer-based conductive materials for improving methane production in anaerobic wastewater treatment systems: a review on their direct interspecific electron transfer mechanism. Fuel 342: 127703. doi: https://doi.org/10.1016/j.fuel.2023.127703.

Tampio, Elina, Ilmari Laaksonen, Karoliina Rimhanen, Niina Honkala, Johanna Laakso, Helena Soinne et al. (2024). Effect of manure co-digestion on methane production, carbon retention, and fertilizer value of digestate. Science of The Total Environment 172083. doi: https://doi.org/10.1016/j.scitotenv.2024.172083.

Tan, Xin-Quan, Wuwei Mo, Abdul Rahman Mohamed and Wee-Jun Ong. (2024). Synergizing aspen plus and life cycle assessment of nascent photocatalytic dry methane reforming over thermocatalytic and biomass gasification toward syngas generation. Journal of Cleaner Production 436: 140270. doi: https://doi.org/10.1016/j.jclepro.2023.140270.

Thalasso, Frederic, Armando Sepulveda-Jauregui, Léa Cabrol, Céline Lavergne, Nazlı Olgun, Karla Martinez-Cruz et al. (2022). Methane and carbon dioxide cycles in lakes of the king george island, maritime Antarctica. Science of The Total Environment 848: 157485. doi: https://doi.org/10.1016/j.scitotenv.2022.157485.

Tian, Hao, Yao Du, Yamin Deng, Xiaoliang Sun, Jiawen Xu, Yiqun Gan and Yanxin Wang. (2024). Identification of methane cycling pathways in quaternary alluvial-lacustrine aquifers using multiple isotope and microbial indicators. Water Research 250: 121027. doi: https://doi.org/10.1016/j.watres.2023.121027.

Vargas-Rodríguez, Jaime E., Angie Cardozo-Núñez, Andrés J. Martínez, Oswan Mora and Juan D. Valderrama-Rincón. (2024). Simultaneous methane production and atmospheric carbon fixation during nutrient recycling from yellow wastewater in a continuously fed PBR-UASB system. Environmental Technology & Innovation 33: 103458. doi: https://doi.org/10.1016/j.eti.2023.103458.

Wang, Jie, Al-attab, K. A. and Teoh Yew Heng. (2024a). Optimization of solid oxide fuel cell system integrated with biomass gasification, solar-assisted carbon capture and methane production. Journal of Cleaner Production 449: 141712. doi: https://doi.org/10.1016/j.jclepro.2024.141712.

Wang, Chenghong, Daming Sun, Qie Shen, Keyi Shen and Yuanyuan Duan. (2024b). Optimization of coalbed methane liquefaction process based on parallel nitrogen reverse brayton cycle under varying methane contents and liquefaction ratios. Energy 293: 130654. doi: https://doi.org/10.1016/j.energy.2024.130654.

Wang, Ling, Chang Liu, Thangavel Sangeetha, Wenzong Liu, Xiaodong Wang, Aijie Wang et al. (2024c). Performance and interaction mechanism of integrating alkali pretreatment and immobilized electrodes on enhanced sludge digestion for methane production. Chemical Engineering Journal 484: 149530. doi: https://doi.org/10.1016/j.cej.2024.149530.

Wang, Xinchu, Biying Chen, Linyong Chen, Guannan Dong, Timothy Csernica, Naizhong Zhang et al. (2024d). Biogenic methane clumped isotope signatures: insights from microbially enhanced coal bed methane. Fuel 365: 131307. doi: https://doi.org/10.1016/j.fuel.2024.131307.

Xu, Aixiang, Lanxiang Yang, Zhiqiang Liu, Yawen Xiong, Xinmeng Xiao, Chengwei Deng and Sheng Yang. (2024). Sustainability evaluation of a PEMFC system for coalbed methane recovery based on life cycle assessment and energy analysis. Journal of Cleaner Production 434: 140344. doi: https://doi.org/10.1016/j.jclepro.2023.140344.

Zhang, Ke, Xiangling Wu, Wei Wang, Jia Chen, Hongbing Luo, Wei Chen et al. (2022). Anaerobic oxidation of methane (AOM) driven by multiple electron acceptors in constructed wetland and the related mechanisms of carbon, nitrogen, sulfur cycles. Chemical Engineering Journal 433: 133663. doi: https://doi.org/10.1016/j.cej.2021.133663.

Zhang, Xinxu, Cuijing Zhang, Yang Liu, Rui Zhang and Meng Li. (2023). Non-negligible roles of archaea in coastal carbon biogeochemical cycling. Trends in Microbiology 31(6): 586–600. doi: https://doi.org/10.1016/j.tim.2022.11.008.

Zhang, Miao, Wenmin Huang, Lei Zhang, Zixuan Feng, Yanxia Zuo, Zuoming Xie and Wei Xing. (2024a). Nitrite-dependent anaerobic methane oxidation (N-DAMO) in global aquatic environments: a review. Science of The Total Environment 921: 171081. doi: https://doi.org/10.1016/j.scitotenv.2024.171081.

Zhang, Yiran, Linjia Li, Shu Zhao, Jiaqi Feng, Ke Li, Zhen Huang et al. (2024b). Effects of Ni loading and Ce doping on a CaO-based dual function material for integrated carbon capture and *in situ* methanation. Catalysis Science & Technology 14(5): 1255–65. doi: https://doi.org/10.1039/d3cy01673k.

Zhou, Chuanqiao, Muchun Zhou, Ruoyu Jia, Yu Peng, Fenjun Zhao, Runze Xu et al. (2023). Particulate organic carbon potentially increases methane emissions from oxic water of eutrophic lakes. Science of The Total Environment 889: 164339. doi: https://doi.org/10.1016/j.scitotenv.2023.164339.

Chapter 5

Methanogens in Extreme Environments

Surviving and Thriving

Prasann Kumar[1,*] and *Joginder Singh*[2]

5.1 Introduction

Methanogens are a taxonomic group of microorganisms classified within Archaea, distinguished by their distinctive capacity to generate methane as a metabolic byproduct via methanogenesis. The organisms above exhibit anaerobic characteristics, indicating their ability to flourish in oxygen-deprived environments (Koul et al. 2021, Welte and Deppenmeier 2014). They are commonly observed in anaerobic habitats, including wetlands, marshes, sediments, and animal gastrointestinal tracts. Methanogens are paramount in the global carbon cycle as they facilitate the conversion of organic matter into methane gas, subsequently enabling its release into the atmosphere. Extreme environments exhibit a range of conditions that are unfriendly to most life forms, encompassing extreme temperatures, pH levels that are either acidic or alkaline, elevated salinity, and reduced oxygen levels. Notwithstanding these obstacles, specific microorganisms, such as methanogens, have developed distinctive adaptations that enable them to flourish in these exceedingly harsh circumstances. Diverse terrestrial and aquatic habitats encompass many extreme environments, including deep-sea hydrothermal vents, acidic hot springs, alkaline lakes, hypersaline environments, and anoxic sediments. In extreme

[1] Department of Agronomy, School of Agriculture, Lovely Professional University, Phagwara, Punjab, 144411, India.
[2] Department of Botany, Nagaland University, Nagaland, India.
* Corresponding author: prasann0659@gmail.com

DOI: 10.1201/9781003506898-5

environments, methanogens are paramount in the biogeochemical cycling of carbon, hydrogen, and various other elements. The production of methane by these organisms results in the emission of greenhouse gas emissions, thereby exerting an influence on global climate patterns (Kumar et al. 2018, Maier and Neilson 2015, Pati et al. 2023, Salwan and Sharma 2020, Santos-Pereira et al. 2024). Examining methanogens in extreme environments yields valuable insights into their ecological functions and their impact on the dynamics of the environment. Methanogens exhibit promising prospects in the field of biotechnology, specifically in the domains of biogas generation and bioremediation. Exploring the mechanisms by which these organisms operate under extreme conditions can enhance biogas reactors' efficiency and devise practical approaches for reducing methane emissions from industrial operations. Extraterrestrial habitats can be better understood by examining extreme environments on Earth, as they offer valuable insights into the possibility of life on other celestial bodies. Methanogens are particularly interested in astrobiology due to their remarkable capacity to endure and thrive in highly challenging environments like those on celestial bodies such as Mars and Europa. Investigating methanogens in extreme environments can contribute to exploring extraterrestrial life and enhance our comprehension of the suitability of various heavenly bodies. To flourish in challenging environments, methanogens have developed adaptive mechanisms such as distinct enzyme systems, membrane compositions, and stress response pathways. By examining these adaptations, researchers can acquire valuable knowledge regarding the evolutionary mechanisms that influence microbial life and the genetic

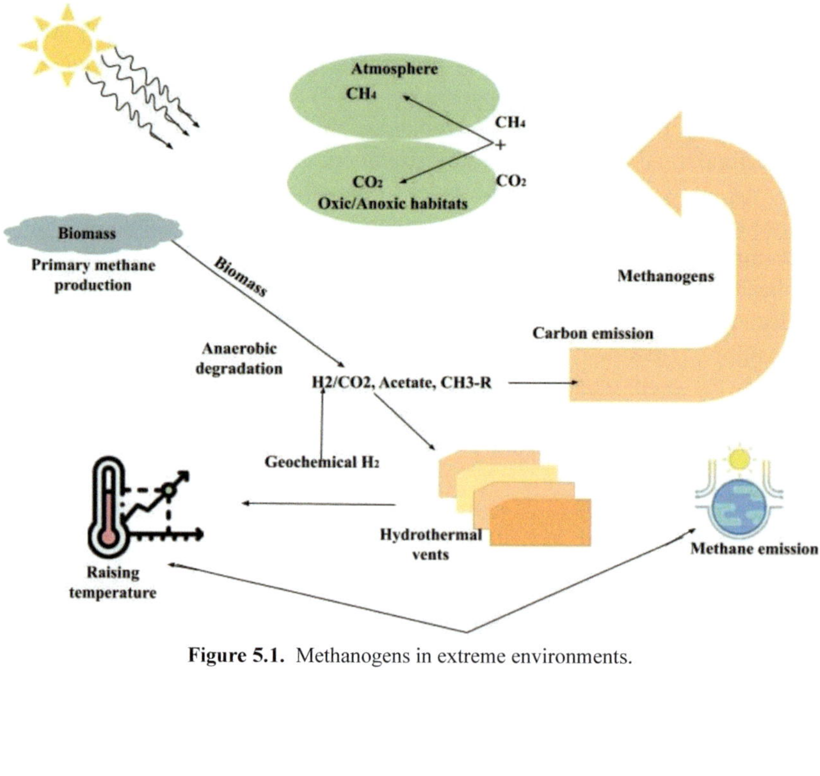

Figure 5.1. Methanogens in extreme environments.

processes that contribute to microbial resilience in extreme environmental conditions. Global warming and climate change are exacerbated by methane emissions from methanogens inhabiting extreme environments. A comprehensive comprehension of the various factors that impact methanogen activity and methane production under extreme conditions is imperative to accurately forecast forthcoming climate scenarios and formulate practical approaches to alleviate greenhouse gas emissions (Castillo and Villafania 2024, Mafra et al. 2022, Peeples 2014). Examining methanogens in exceptionally challenging environments yields significant knowledge in microbial ecology, biogeochemistry, biotechnology, astrobiology, and climate science. By deciphering the enigmatic nature of these extraordinary microorganisms, researchers have the potential to unveil novel prospects for the preservation of the environment, advancements in technology, and the exploration of outer space (Fig. 5.1).

5.2 Diversity of Extreme Environments Inhabited by Methanogens

Methanogens are exceptional microorganisms that can flourish in various harsh environments, showcasing their ability to adapt to conditions that are inhospitable to most organisms. The wide range of habitats in which methanogens are found enables them to assume significant roles in diverse biogeochemical processes, making substantial contributions to global methane production and influencing ecological dynamics. This article examines the wide range of extreme environments inhabited by methanogens, encompassing hydrothermal vents with high temperatures and anoxic sediments with low oxygen levels. Hydrothermal vents near the ocean floor, where tectonic plates separate and seawater infiltrates the Earth's crust (Jadhav et al. 2022, Maslać et al. 2022, Passarini et al. 2022, Shrestha et al. 2019). Magma chambers then heat it and then release it back into the ocean. The vents release high-temperature fluids rich in minerals and contain hydrogen and carbon dioxide, creating optimal circumstances for the proliferation of methanogens. Methanogens within hydrothermal vent systems significantly generate methane by utilising hydrogen as an electron donor to reduce carbon dioxide. Geothermal springs are terrestrial thermal springs that derive energy from geothermal heat from the Earth's mantle. These springs are commonly located in volcanic regions or geologically active areas. The temperatures of these springs can exceed the boiling point of water and are distinguished by either acidic or alkaline conditions, which are contingent upon the geological composition of the adjacent rocks. Methanogens in geothermal springs have evolved to thrive in environments characterised by elevated temperatures and extreme pH levels. These organisms play crucial roles in the production of methane and the cycling of carbon within these distinct ecological systems. Acid Mine Drainage (AMD) results from the exposure of sulphide minerals in rocks to air and water, forming sulphuric acid. AMD environments exhibit high acidity and frequently harbour heavy metal contamination, presenting substantial obstacles for most life forms. Despite the challenging environmental circumstances, methanogens have been detected in sites affected by AMD, where they generate methane via anaerobic fermentation mechanisms. Geothermal features known as acidic hot springs, acid geysers, or fumaroles are distinguished by their acidic pH levels and elevated temperatures. The springs in question are comm only linked to volcanic phenomena and exhibit a high concentration of sulphuric compounds and various other minerals. In acidic hot springs, methanogens adapt to cope with low pH conditions, facilitating methane production through metabolic pathways that differ from those observed in neutral or alkaline environments. Alkaline lakes, alternatively referred to as soda lakes, are aquatic environments characterised by elevated pH levels, typically surpassing, primarily attributed to the presence of dissolved minerals, including carbonates and bicarbonates (Chastain and Kral 2010, Chaudhary et al. 2022, Coolen et al. 2004, Lam et al. 2020). These lakes are frequently observed in arid regions with restricted freshwater input and elevated evaporation rates. Methanogens inhabiting alkaline lakes have evolved to thrive in highly acidic environments, enabling them to generate methane through fermenting organic substances without oxygen. The subtype of alkaline lakes known as soda lakes is distinguished by their elevated levels of dissolved sodium carbonate. These lakes provide a habitat for distinct microbial communities that have adapted to the elevated salinity and alkaline pH conditions. Among these communities are methanogens, which significantly affect methane production within these environments. Salt flats, alternatively referred to as salt pans or salt deserts, are expansive landforms enveloped by salt crusts resulting from water evaporation from saline lakes or seas. These environments are characterised by extreme aridity and high salinity levels, which make them unfriendly to most life forms. Methanogens inhabiting salt flats can endure hypersaline environments and actively participate in the generation of methane using the anaerobic metabolism of organic substances. Hydrosaline environments, such as salt marshes and saltwater lagoons, are characterised by water with salinity levels surpassing seawater. Halophilic microorganisms, such as methanogens adapted to high salt concentrations, frequently inhabit these environments. These microorganisms are involved in carbon cycling and methane production. Anaerobic sediments with restricted or non existent oxygen availability are commonly observed in aquatic habitats, including lakes, rivers, and marine basins. The sediments exhibit a high concentration of organic matter, creating favourable circumstances for anaerobic microbial processes, such as methane generation by methanogens. Soils depleted of oxygen, such as wetlands and marshes, undergo periodic or permanent saturation with water, resulting in anaerobic atmospheric conditions (Gupta et al. 2019, Jabbari et al. 2024, Johnson and Aguilera 2019, Thompson et al. 2022). Methanogens in soils with reduced oxygen levels significantly release methane through organic matter decomposition in anaerobic environments. These organisms are essential contributors to carbon cycling within these ecological systems. Methanogens demonstrate exceptional versatility in adapting to harsh environments, spanning from heated hydrothermal vents to oxygen-deprived sediments with low oxygen levels. Methanogens play crucial roles in biogeochemical cycling and ecological functioning by colonising these habitats and contributing to methane production, thereby underscoring their significance in the global carbon cycle. Comprehending the habitats occupied by methanogens offers valuable insights into their ecological functions and ability to adapt to harsh conditions. This knowledge has significant implications for environmental conservation, biotechnology, and astrobiology (Fig. 5.2).

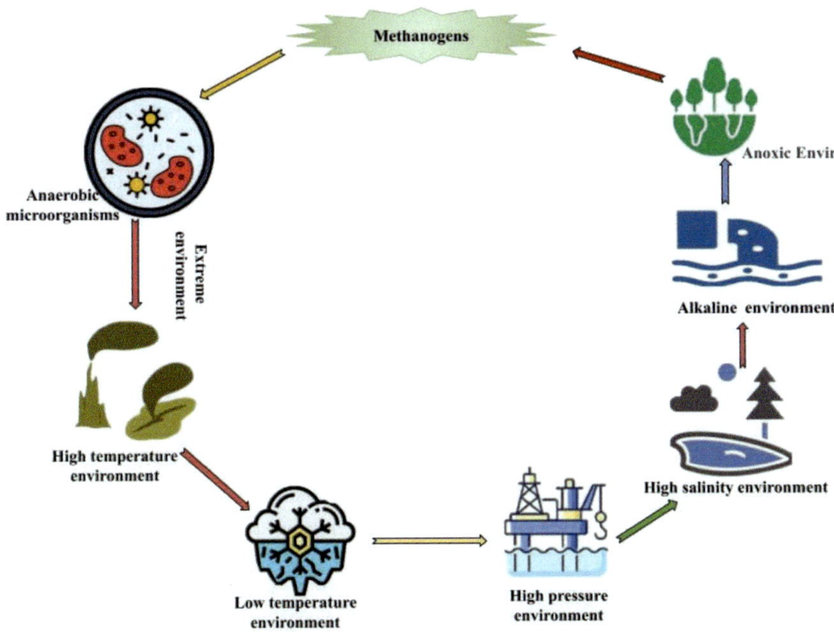

Figure 5.2. Diversity of extreme environments inhabited by methanogens.

5.3 Metabolic Pathways of Methanogens in Extreme Environments

The methanogens, which encompass a wide range of archaea, are widely recognised for their distinctive capacity to generate methane as a byproduct of their metabolic processes. Despite the challenging circumstances in extreme environments, methanogens have developed complex metabolic pathways to flourish and contribute substantially to biogeochemical cycles. This article examines the metabolic pathways of methanogens, specifically emphasising hydrogenotrophic methanogenesis, acetoclastic methanogenesis, and methylotrophic methanogenesis. The objective is to investigate how these pathways facilitate the survival and flourishing of methanogens in challenging ecological conditions. Hydrogenotrophic methanogenesis represents a fundamental metabolic pathway employed by methanogens for methane production. Methanogens employ molecular hydrogen (H_2) as the electron donor and carbon dioxide (CO_2) as the terminal electron acceptor in this particular pathway to generate methane (CH_4). Hydrogenotrophic methanogens engage in diverse metabolic pathways to generate hydrogen, such as the fermentation of organic compounds by syntrophic bacteria or the decomposition of intricate organic substances in anaerobic settings. Methanogens contain hydrogenases, enzymes that facilitate the reversible transformation of molecular hydrogen into protons and electrons. In hydrogen oxidation, methanogens employ the electrons produced to convert carbon dioxide into formate (HCOO-) using formate dehydrogenase enzymes.

Formyl-methanofuran dehydrogenase facilitates the conversion of formate into methylene-tetrahydromethanopterin (CH2-THMPT), an intermediate compound in methanogenesis (Maier 2009, Mehta et al. 2021, Nagarajan et al. 2022, Rogers 2015, Singh and Porwal 2022, Sinha et al. 2017). The enzyme methyl-tetrahydromethanopterin reductase catalyses methylene-tetrahydromethanopterin reduction to methyl-tetrahydromethanopterin (CH3-THMPT). In the concluding methanogenesis stage, methyl-tetrahydromethanopterin serves as a methyl group donor. This donor is subsequently transferred to coenzyme M (CoM) to generate methane and coenzyme B (CoB) through the catalytic activity of Methyl-Coenzyme M Reductase (MCR), a pivotal enzyme involved in the process of methanogenesis. Hydrogenotrophic methanogenesis involves energy conservation by methanogens by generating a proton motive force across the cell membrane. This motive force subsequently facilitates the synthesis of adenosine triphosphate (ATP) via the action of ATP synthase enzymes. Methanogens residing in extreme environments with restricted organic carbon availability and elevated hydrogen concentrations, such as hydrothermal vents and deep-sea sediments, rely heavily on hydrogenotrophic methanogenesis as a vital metabolic pathway. Methanogens contribute to methane production and play crucial roles in carbon cycling in these ecosystems by using hydrogen and carbon dioxide as substrates. Certain methanogens utilise acetoclastic methanogenesis as an alternative metabolic pathway to generate methane from acetate, a shared organic compound in anaerobic conditions. The process of acetate activation results in the formation of acetyl-CoA, which functions as the principal substrate for methanogenesis. The activation of acetate to acetyl-CoA occurs via the catalytic activity of acetate kinase and phosphotransacetylase enzymes. These enzymes catalyse the phosphorylation of acetate, resulting in the formation of acetyl-phosphate. Subsequently, the phosphate group is transferred to coenzyme A (CoA), producing acetyl-CoA. In acetoclastic methanogenesis, acetyl-CoA acts as the methyl group donor and is transformed into methyl-tetrahydromethanopterin (CH3-THMPT) through the action of acetyl-CoA decarbonylase/synthase enzymes. In the concluding stage of methanogenesis, methyl-tetrahydromethanopterin is employed and subsequently transferred to coenzyme M (CoM) for the synthesis of methane and coenzyme B (CoB) through the catalytic activity of methyl-coenzyme M reductase (MCR) enzymes. Acetate-rich environments, such as anaerobic sediments, wastewater treatment facilities, and rice paddies, are a significant source of acetoclastic methanogenesis. Methanogens that can produce methane through acetoclastic processes are crucial for producing methane and cycling carbon in these environments. They contribute significantly to greenhouse gas emissions and the dynamics of the ecosystem. The process of methylotrophic methanogenesis is a metabolic pathway employed by certain methanogens to generate methane through the utilisation of methylated compounds, including methanol, methylamines, and methylated sulphides (Besse et al. 2015, Fenchel et al. 2012, Joseph 2023, Margesin and Miteva 2011, Mohanty et al. 2022, Shao and Whitman 2023). Methylotrophic methanogenesis facilitates the utilisation of a more comprehensive array of substrates by methanogens for methane production, thereby augmenting their ecological niche in highly challenging environments. Certain enzymes, including methyltransferases,

Table 5.1. Different types of metabolic pathways in methanogens (Source: Chastain and Kral 2010, Chaudhary et al. 2022, Coolen et al. 2004, Lam et al. 2020).

Metabolic Pathway	**Description**
CO_2 Reduction Pathway	Utilises CO_2 as a carbon source, often through the Wood-Ljungdahl pathway.
Acetate Production	Conversion of intermediates like H_2 and CO_2 to acetate through various pathways.
Methanogenesis	Production of methane from substrates such as acetate, methylamines, or methanol.
Methylotrophic Pathway	Methanogens are capable of utilising methylated compounds like methanol for methanogenesis.
H_2-Dependent Pathway	Utilisation of molecular hydrogen (H_2) as an electron donor for methanogenesis.
CO-Dependent Pathway	Utilisation of carbon monoxide (CO) as a substrate for methanogenesis.
Sulphate Reduction	Some methanogens can compete with sulphate-reducing bacteria by reducing sulphate.
Nitrogen Fixation	Capability to fix atmospheric nitrogen for incorporation into cellular components.
Energy Generation	Utilisation of proton gradients for ATP synthesis, often via ATP synthase.
Glycolysis	Breakdown of glucose to pyruvate, yielding ATP and precursor molecules.
Pentose Phosphate Pathway	Produces NADPH and ribose-5-phosphate for nucleotide synthesis.
Pyruvate Fermentation	Convert pyruvate to various end products like lactate, ethanol, or acetate.
Lipid Metabolism	Biosynthesis and degradation of lipids for membrane formation and energy storage.
Amino Acid Metabolism	Utilisation of amino acids as carbon and nitrogen sources for cellular processes.
DNA Repair	Mechanisms to repair damaged DNA are crucial for survival in extreme environments.
Stress Response	Activation of pathways to mitigate oxidative, osmotic, or thermal stress.
Protein Folding	Chaperone proteins assist in the proper folding of proteins under extreme conditions.
Iron-Sulphur Cluster Biosynthesis	Synthesis of iron-sulphur clusters, essential cofactors for various enzymes.
Phosphate Metabolism	The acquisition, utilisation, and recycling of phosphate are critical cellular components.
Carbon Fixation	Ability to fix inorganic carbon (CO_2) into organic molecules via various pathways.
Nitrogen Metabolism	Assimilation and cycling of nitrogen-containing compounds within the cell.

Table 5.1 contd. ...

...Table 5.1 contd.

Metabolic Pathway	Description
Lipopolysaccharide Biosynthesis	Production of lipopolysaccharides for cell membrane stability and protection.
Metal Ion Homeostasis	Maintenance of appropriate concentrations of metal ions for cellular functions.
Polyhydroxyalkanoate (PHA) Biosynthesis	Production of PHA for carbon storage under nutrient-limiting conditions.
Quorum Sensing	Communication system regulating gene expression in response to population density.
Virulence Factor Production	Synthesis of molecules aiding in host colonisation or evasion of host defences.
Biofilm Formation	Formation of structured microbial communities for protection and resource sharing.
CRISPR-Cas System	The adaptive immune system for defence against invading genetic elements.
Toxin-Antitoxin Systems	Regulation of cell growth and survival under stress conditions through toxin release.
Secondary Metabolite Biosynthesis	The production of compounds is not directly involved in growth, and it is often used for defence.

methylamine dehydrogenases, and methanol-cobalamin methyltransferases, are present in methylotrophic methanogens. These enzymes facilitate the use of methylated compounds as substrates for methanogenesis. The demethylation of methylated compounds results in the formation of methyl-tetrahydromethanopterin (CH3-THMPT), which functions as the methyl group donor during methanogenesis. In the last stage of methanogenesis, methyl-tetrahydromethanopterin is used to transfer methane and coenzyme B (CoB) to coenzyme M (CoM) through the action of methyl-coenzyme M reductase (MCR) enzymes. The process of methylotrophic methanogenesis facilitates the flourishing of methanogens in habitats characterised by a high concentration of methylated compounds, including marine sediments, anaerobic soils, and animal gastrointestinal tracts. Methylotrophic methanogens play a significant role in carbon cycling and ecological processes in extreme environments by utilising a wide range of substrates to produce methane. A diverse range of metabolic pathways, such as hydrogenotrophic, acetoclastic, and methylotrophic methanogenesis, are utilised by methanogens to generate methane and flourish in challenging ecological conditions. The metabolic pathways employed by methanogens facilitate their utilisation of various substrates and their ability to adapt to diverse environmental niches (Ciaramella et al. 2005, Dion 2023, Dopffel et al. 2024, Kochhar et al. 2022, Li et al. 2022, Liu et al. 2016, Moissl-Eichinger et al. 2018, Zhou et al. 2024). These pathways play pivotal roles in carbon cycling, greenhouse gas emissions, and overall dynamics within the ecosystem. Gaining a comprehensive understanding of the metabolic adaptability exhibited by methanogens offers valuable insights into their environmental importance and evolutionary responses to challenging circumstances. These findings have significant implications for various fields, including biogeochemical cycling, ecological preservation, and biotechnological utilisation (Table 5.1).

5.4 Stress Response Mechanisms of Methanogens

Methanogens, as extremophiles, thrive in environments characterised by extreme conditions such as high temperature, salinity, pH variations, and oxidative stress. To survive and maintain cellular integrity under such challenging conditions, methanogens have evolved sophisticated stress response mechanisms. In this article, we explore the stress response mechanisms employed by methanogens, including heat shock proteins, osmoprotectants, pH homeostasis mechanisms, and oxidative stress defence mechanisms. Heat Shock Proteins (HSPs) play a crucial role in the stress response of methanogens to high temperatures. These proteins are molecular chaperones that assist in the folding, assembling, and degrading of other proteins, thereby protecting them from thermal denaturation and aggregation. Methanogens upregulate the expression of heat shock proteins, such as Hsp60 (chaperonin) and Hsp70 (DnaK), in response to elevated temperatures. The transcriptional activation of heat shock genes is mediated by heat shock transcription factors (HSFs), which bind to Heat Shock Elements (HSEs) in the promoter regions of target genes. Heat shock proteins facilitate the refolding of denatured proteins and prevent their irreversible aggregation under heat-stress conditions. Chaperonins, such as Hsp60, provide a protected environment for protein folding, while Hsp70 assists in the disaggregation and refolding of misfolded proteins (Alpana et al. 2017, Bonfanti et al. 2018, Cena et al. 2024, Gast et al. 2009, Gupta et al. 2023, Holmes and Smith 2016, Nie et al. 2021, Tian et al. 2018). In severe heat stress conditions, irreversibly damaged proteins may be targeted for proteolytic degradation by ATP-dependent proteases, such as Lon protease and Clp protease, to prevent the accumulation of toxic protein aggregates. The induction of heat shock proteins and their role in protein folding and repair are essential for maintaining cellular homeostasis and viability in thermophilic methanogens inhabiting high-temperature environments, such as hydrothermal vents and geothermal springs. Methanogens often encounter osmotic stress in environments characterised by high salinity or fluctuations in osmotic pressure. To counteract the harmful effects of osmotic stress, methanogens employ osmoprotectants, which are small organic molecules that help maintain cellular osmotic balance and stabilise macromolecular structures. Glycine betaine is a common osmoprotectant synthesised by methanogens in response to osmotic stress. Methanogens produce glycine betaine from choline via the enzymatic action of choline oxidase and betaine aldehyde dehydrogenase. Proline is another osmoprotectant accumulated by methanogens to counteract osmotic stress. Methanogens can synthesise proline from glutamate through a series of enzymatic reactions involving pyrroline-5-carboxylate reductase and ornithine cyclodeaminase. Some methanogens accumulate trehalose, a disaccharide sugar, as an osmoprotectant. Trehalose stabilises proteins and membranes under osmotic stress conditions and protects cellular structures from dehydration. The accumulation of osmoprotectants helps methanogens maintain cellular hydration and structural integrity in high-salinity environments, such as salt flats and hypersaline lakes, enabling them to thrive under osmotically challenging conditions. Methanogens inhabit environments with a wide range of pH Values, including acidic hot springs and alkaline soda lakes. To maintain intracellular pH homeostasis and withstand pH fluctuations, methanogens employ various mechanisms, including actively pumping protons out of the cell or sequestering them in intracellular compartments to regulate cytoplasmic pH. Proton-translocating ATPases and proton-pumping membrane proteins contribute to proton extrusion or sequestration in response to changes in extracellular pH. Methanogens possess intracellular buffering systems, such as the bicarbonate/carbon dioxide (HCO_3^-/CO_2) system, to maintain stable pH conditions within the cell. Bicarbonate ions act as a buffer, absorbing excess protons and stabilising intracellular pH in response to acidic stress. Methanogens regulate the expression and activity of membrane transporters involved in ion transport and pH regulation. pH-responsive transcriptional regulators modulate the expression of genes encoding transporters and enzymes involved in pH homeostasis (Besse et al. 2015, Fenchel et al. 2012, Johnson and Aguilera 2019, Joseph 2023, Liu et al. 2016, Margesin and Miteva 2011, Mehta et al. 2021, Mohanty et al. 2022, Shao and Whitman 2023). Methanogens can thrive in environments with extreme pH conditions by employing pH homeostasis mechanisms, contributing to biogeochemical cycling and ecosystem functioning in acidic and alkaline habitats. Although methanogens are typically anaerobic microorganisms, they may encounter oxidative stress in environments where oxygen is present or during exposure to Reactive Oxygen Species (ROS). To mitigate oxidative damage and maintain redox balance, methanogens employ oxidative stress defence mechanisms, including methanogens produce antioxidant enzymes, such as superoxide dismutase (SOD), catalase, and peroxidases, to detoxify reactive oxygen species and prevent oxidative damage to cellular components. SOD catalyses the dismutation of superoxide radicals (O_2^-) to hydrogen peroxide (H_2O_2), which is further detoxified by catalase and peroxidases. Methanogens regulate cellular redox balance by maintaining a reduced intracellular environment through the action of redox-active molecules and cofactors, such as reduced ferredoxin and coenzyme F420. Electron transport chains associated with methanogenesis enzymes also contribute to redox balance maintenance by consuming reducing equivalents and preventing the accumulation of reactive oxygen species. Methanogens possess DNA repair systems, including base excision repair and homologous recombination pathways, to repair oxidative DNA damage caused by reactive oxygen species. DNA repair enzymes, such as DNA glycosylases and DNA ligases, facilitate removing and repairing oxidised DNA lesions to maintain genomic integrity. By deploying oxidative stress defence mechanisms, methanogens can mitigate the harmful effects of oxidative stress and survive in environments where oxygen exposure occurs transiently or sporadically, such as oxygenated sediments and microbial mats (Castillo and Villafania 2024, Chaudhary et al. 2022, Jadhav et al. 2022, Lam et al. 2020, Mafra et al. 2022, Maslać et al. 2022, Passarini et al. 2022, Peeples 2014, Salwan and Sharma 2020, Santos-Pereira et al. 2024, Shrestha et al. 2019). Methanogens employ stress response

Table 5.2. Stress response mechanism in methanogens (Source: Chastain and Kral 2010, Chaudhary et al. 2022, Coolen et al. 2004, Jadhav et al. 2022, Lam et al. 2020, Maslać et al. 2022, Pati et al. 2023, Shrestha et al. 2019).

Stress Response Mechanism	Description
Heat Shock Proteins	Synthesis of chaperone proteins to assist in protein folding and prevent protein denaturation.
Cold Shock Proteins	Production of proteins that maintain membrane fluidity and prevent cell damage at low temperatures.
Osmotic Stress Response	Activation of osmoregulatory mechanisms to maintain cellular turgor pressure and prevent cell lysis.
Oxidative Stress Defence	Synthesis of antioxidants like superoxide dismutase and catalase to mitigate oxidative damage.
DNA Repair Mechanisms	Various repair pathways (e.g., base excision repair, nucleotide excision repair) are activated to fix DNA damage.
Protein Degradation Systems	Upregulation of proteases and degradation machinery to remove misfolded or damaged proteins.
Metal Ion Homeostasis Control	Regulation of intracellular metal ion concentrations to prevent metal toxicity or deficiency.
Lipid Composition Adjustment	Alteration of membrane lipid composition to maintain membrane fluidity under stress conditions.
Compatible Solutes Synthesis	Synthesis and accumulation of compatible solutes (e.g., betaine, proline) to counteract osmotic stress.
Cryoprotectant Synthesis	Cryoprotectants (e.g., glycerol) are produced to prevent cellular damage from freezing temperatures.
Biofilm Formation	Formation of biofilms for protection against environmental stresses and resource sharing.
Quorum Sensing Systems	Communication systems regulate gene expression in response to cell density to coordinate stress responses.
Toxin-Antitoxin Systems	Regulation of cell growth and survival through toxin production and subsequent antitoxin activity.
RNA Thermometers	Regulation of gene expression by temperature-dependent RNA structures affecting translation efficiency.
Ion Channels Regulation	Control ion channel activity to maintain ion homeostasis and membrane potential under stress.
Efflux Pump Activation	Upregulation of efflux pumps to remove toxic compounds or metabolites from the cell.
Secondary Messenger Systems	Activation of signalling pathways mediated by secondary messengers (e.g., cyclic AMP, c-di-GMP).
Redox Homeostasis Maintenance	Maintenance of cellular redox balance through regulation of redox cofactor levels and metabolism.
Virulence Factor Regulation	Control of virulence factor expression to modulate interactions with host organisms.
Carbon Storage Mobilization	Stored carbon reserves (e.g., glycogen, polyhydroxyalkanoates) are mobilised under stress conditions.

Table 5.2 contd. ...

...Table 5.2 contd.

Stress Response Mechanism	Description
RNA Stability Control	Stabilisation of mRNA transcripts or regulation of RNA degradation rates in response to stress.
Antioxidant Enzyme Activation	Activation of enzymes such as peroxidases and reductases to scavenge reactive oxygen species.
Transcriptional Regulators	Modulation of gene expression through transcription factors or alternative sigma factors.
Phage Defence Mechanisms	Strategies to resist or counteract phage infection, including restriction-modification systems.
Spore Formation	Formation of spores as a survival strategy under unfavourable conditions.
Cell Cycle Arrest	Arrest of the cell cycle to prevent proliferation and conserve cellular resources under stress.
Ribosome Regulation	Regulation of ribosome biogenesis and activity to adapt to changing growth conditions.
Cell Wall Remodelling	Modification of cell wall composition or structure to withstand environmental stresses.
Enzyme Inhibition	Inhibition of specific enzymes to regulate metabolic pathways in response to stress signals.
Secretion Systems Activation	Activation of secretion systems to export proteins or toxins in response to stress.
Phospholipid Bilayer Repair	Repair of damaged phospholipid bilayers through lipid remodelling or vesicle fusion.
ATP Synthesis Adjustment	Adjustment of ATP synthesis rates to meet cellular energy demands under stress conditions.
Cyclic di-AMP/di-GMP Signalling	Regulation of cellular processes through cyclic dinucleotide signalling molecules.
Histone-Like Proteins	Use of histone-like proteins to compact DNA or regulate gene expression in response to stress.
RNA Interference Mechanisms	Regulation of gene expression at the post-transcriptional level through RNA interference pathways.
Protein Folding Assistance	Assistance from chaperone proteins in proper folding and assembly of proteins under stress.
pH Homeostasis Maintenance	Maintenance of intracellular pH through ion transport and buffering systems.

mechanisms, including heat shock proteins, osmoprotectants, pH homeostasis, and oxidative stress defence mechanisms, to adapt to extreme environmental conditions and maintain cellular viability. Understanding the molecular basis of stress adaptation in methanogens provides insights into their ecological resilience and evolutionary adaptation to diverse habitats, with implications for biogeochemical cycling, environmental sustainability, and biotechnological applications (Table 5.2).

5.5 Adaptations of Methanogens to Extreme Conditions

Methanogens, a heterogeneous assemblage of microorganisms classified under the taxonomic group Archaea, demonstrate notable adaptations that facilitate their successful survival in environments characterised by elevated temperatures, salinity, pH extremes, and various other demanding circumstances. The adaptations above play a vital role in ensuring living organisms' survival and metabolic functioning in multiple environments, including hydrothermal vents, hypersaline lakes, and acidic hot springs. The present article investigates methanogens' essential adaptations in response to extreme environmental conditions. These adaptations include membrane composition and stability, DNA repair mechanisms, enzyme responses, and metabolic flexibility. Methanogenic organisms exhibit distinctive membrane lipids called ether lipids, comprising isoprenoid chains connected to glycerol through ether bonds. Compared to the ester-linked lipids in bacteria and eukaryotes, ether lipids exhibit superior stability and resistance to hydrolysis, rendering them highly suitable for extreme environments. Certain methanogens, especially those residing in high-temperature habitats like hydrothermal vents, possess monolayer membranes consisting of stacked lipid monolayers. Monolayer membranes provide thermal stability and prevent disruption of lipid bilayers in the presence of high temperatures, thereby enabling methanogens to preserve their membranes' integrity. Some thermophilic methanogens can synthesise tetraether lipids, forming stable lipid monolayers or bilayers and resisting extreme temperatures and pH levels (Besse et al. 2015, Ciaramella et al. 2005, Dion 2023, Dopffel et al. 2024, Fenchel et al. 2012, Joseph 2023, Li et al. 2022, Liu et al. 2016, Margesin and Miteva 2011, Mohanty et al. 2022, Moissl-Eichinger et al. 2018, Shao and Whitman 2023, Singh and Porwal 2022). Tetraether lipids are characterised by cyclopentane rings, branched chains, and hydroxyl groups, contributing to membranes' structural integrity and durability in challenging conditions. To rectify DNA damage resulting from exposure to severe conditions such as oxidative stress and UV radiation, methanogens utilise base excision repair pathways. Base Excision Repair (BER) is a process that entails the elimination of impaired bases by DNA glycosylases, which is then followed by the activity of DNA polymerases and ligases to substitute and secure the infected DNA strand. Methanogens employ homologous recombination as a mechanism to rectify double-strand breaks and other intricate DNA lesions that arise as a result of extreme environmental circumstances. Recombination (HR) is a process that entails the transfer of genetic material between homologous DNA sequences, facilitated by recombinases and DNA repair proteins, to restore the integrity of the genome. Thermophilic methanogens can synthesise heat-stable enzymes that have evolved to accommodate elevated temperatures, thereby augmenting metabolic activity and enzyme efficacy under extreme heat conditions. Thermophilic enzymes demonstrate structural modifications, including heightened hydrophobic interactions and salt bridges, to sustain stability and functionality under high temperatures. In hypersaline environments, methanogens can synthesise specifically adapted enzymes to operate effectively under high salt concentrations, thereby mitigating the risk of enzyme denaturation and inactivation. Enzymes tolerant to salt may exhibit surface residues with negative charges and osmoprotectant-binding sites, which protect the enzyme

Table 5.3. Different types of adaptations of methanogens to extreme conditions (Source: Castillo and Villafania 2024, Chastain and Kral 2010, Chaudhary et al. 2022, Coolen et al. 2004, Gupta et al. 2019, Jadhav et al. 2022, Lam et al. 2020, Mafra et al. 2022, Maslać et al. 2022, Passarini et al. 2022, Peeples 2014, Salwan and Sharma 2020, Shrestha et al. 2019, Thompson et al. 2022).

Adaptation	**Description**
Psychrophilic Adaptation	Adaptation to cold temperatures, including adjustments in membrane fluidity and enzyme activity.
Thermophilic Adaptation	Tolerance to high temperatures, often involving thermostable enzymes and heat shock response systems.
Halophilic Adaptation	Ability to thrive in high salt environments involving osmoregulation and compatible solute synthesis.
Acidophilic Adaptation	Adaptation to acidic conditions, often involving pH homeostasis mechanisms and acid-stable enzymes.
Alkaliphilic Adaptation	Tolerance to alkaline conditions, including adjustments in proton and ion transport systems.
Barophilic Adaptation	Adaptation to high-pressure environments often involves adjustments in membrane structure and composition.
Metal Tolerance	Tolerance to high metal concentrations involving metal ion transport and detoxification mechanisms.
Desiccation Resistance	Resistance to dehydration or desiccation, often involving the production of protective exopolysaccharides.
UV Radiation Tolerance	Ability to withstand ultraviolet (UV) radiation, involving DNA repair mechanisms and sunscreen pigments.
Osmotic Stress Tolerance	Tolerance to changes in osmotic pressure involving compatible solute synthesis and ion transport.
Redox Balance Regulation	Maintenance of redox balance under extreme conditions through adjustments in electron transport systems.
Anaerobic Respiration	Utilisation of alternative electron acceptors without oxygen, such as nitrate or sulphate.
Aerotolerance	Tolerance to oxygen exposure, often involving enzymes like superoxide dismutase and catalase.
Dormancy	Ability to enter a dormant state under unfavourable conditions, conserving energy, and resources.
Horizontal Gene Transfer	Acquisition of genes from other organisms to adapt to changing environmental conditions.
Biofilm Formation	Formation of biofilms for protection and resource sharing in nutrient-poor or harsh environments.
Antioxidant Enzyme Production	Synthesis of enzymes to scavenge reactive oxygen species and mitigate oxidative stress.
Methanogenesis Regulation	Regulation of methanogenesis rates to match substrate availability and energy demands.
DNA Repair Mechanisms	Activation of repair pathways to fix DNA damage caused by environmental stresses.
Cell Wall Modifications	Modifications in cell wall structure to withstand osmotic stress or changes in pH.

Table 5.3 contd. ...

...Table 5.3 contd.

Adaptation	**Description**
Membrane Permeability Adjustments	Adjustments in membrane permeability to control ion and solute transport under extreme conditions.
Nutrient Storage	Accumulation of nutrient reserves (e.g., glycogen, polyhydroxyalkanoates) for use during scarcity.
pH Homeostasis Mechanisms	Maintenance of intracellular pH through proton transport and buffering systems.
Spore Formation	Formation of spores as a survival strategy under adverse conditions triggered by nutrient limitation or stress.
Cyclic-di-GMP Signalling	Regulation of cellular processes through cyclic dinucleotide signalling molecules.
Carbon Utilisation Strategies	Efficient utilisation of diverse carbon sources for energy production and biosynthesis.
Nitrogen Fixation	Ability to fix atmospheric nitrogen for incorporation into cellular components.
Metal Ion Chelation	Binding of metal ions by chelators or metallothioneins to prevent toxicity or maintain homeostasis.
Quorum Sensing Systems	Communication systems regulate gene expression in response to cell density or environmental cues.
Ribosome Modification	Modifications in ribosome structure or composition to adapt to extreme temperatures or pressures.
RNA Thermometers	Regulation of gene expression by temperature-dependent RNA structures affecting translation efficiency.
Virulence Factor Regulation	Control of virulence factor expression to modulate interactions with host organisms.
Secretion Systems Activation	Activation of secretion systems to export proteins, toxins, or efflux of harmful compounds.
Phospholipid Bilayer Remodelling	Remodelling of phospholipid bilayers to maintain membrane integrity under extreme conditions.
RNA Interference Mechanisms	Regulation of gene expression at the post-transcriptional level through RNA interference pathways.
Protein Folding Assistance	Assistance from chaperone proteins in proper folding and assembly of proteins under stress conditions.
CRISPR-Cas System	The adaptive immune system is used to defend against invading genetic elements, including phages and plasmids.
Cell Aggregation	Formation of aggregates or clusters for protection against environmental stresses.
S-layer Production	Production of surface layer proteins for protection against osmotic, pH, or temperature stresses.
Gas Vesicle Formation	Formation of gas-filled vesicles for buoyancy regulation in stratified environments.
Flagella Regulation	Regulation of flagella synthesis and motility in response to environmental cues or stresses.

from destabilisation caused by salt exposure. Methanogens demonstrate metabolic adaptability by utilising diverse substrates for methanogenesis, encompassing hydrogen, carbon dioxide, acetate, methylamines, and methanol. The utilisation of various substrates enables methanogens to effectively respond to variations in nutrient availability and environmental conditions, thereby ensuring the long-term viability of their metabolic processes. In the process of methanogenesis, methanogens utilise energy conservation mechanisms, including chemiosmotic coupling and electron bifurcation, to produce ATP and decrease the number of electron carriers. The method of chemiosmotic coupling involves the transfer of protons across membranes to facilitate the synthesis of ATP. On the other hand, electron bifurcation serves to conserve energy by combining exergonic and endergonic redox reactions. To flourish in harsh environments, methanogens have developed a range of adaptations, such as distinctive membrane compositions, resilient DNA repair mechanisms, enzyme adaptations, and metabolic flexibility (Bonfanti et al. 2018, Ciaramella et al. 2005, Dion 2023, Dopffel et al. 2024, Gast et al. 2009, Gupta et al. 2023, Kochhar et al. 2022, Li et al. 2022, Tian et al. 2018, Zhou et al. 2024). Methanogens possess these adaptations that allow them to preserve the structure of their cells, maintain their metabolic processes, and inhabit a wide range of environments, including deep-sea hydrothermal vents and terrestrial hot springs. Insights into the ecological success of methanogens in extreme environments and their potential applications in biotechnology and bioremediation efforts can be gained by comprehending the molecular basis of these adaptations (Table 5.3).

5.6 Ecological Significance of Methanogens in Extreme Environments

Methanogens, microorganisms classified under the taxonomic domain Archaea, have a pivotal role in the ecological dynamics of extreme environments. They make substantial contributions to carbon cycling, methane generation and utilisation, and the organisation and operation of microbial communities. This review thoroughly examines the ecological importance of methanogens in wild environments, encompassing their involvement in carbon cycling, methane dynamics, and interactions with other microorganisms. Methanogens play a crucial role in carbon cycling within extreme environments characterised by the limited availability of organic carbon. Methanogens engage in the metabolic process of organic matter within anaerobic environments, such as hydrothermal vents and acidic hot springs. This metabolic activity produces methane as a byproduct, facilitating carbon fixation and recycling (Fenchel et al. 2012, Gupta et al. 2019, Jabbari et al. 2024, Johnson and Aguilera 2019, Maier 2009, Mehta et al. 2021, Nagarajan et al. 2022, Rogers 2015, Sinha et al. 2017, Thompson et al. 2022). Certain methanogens employ carbon dioxide as a carbon source for the process of methanogenesis via the reductive acetyl-CoA pathway, also known as the Wood-Ljungdahl pathway. Methanogens play a crucial role in carbon fixation, preserving carbon equilibrium in extreme environments and supporting microbial communities in nutrient-deficient ecosystems. Methanogens serve as the predominant sources of methane in anaerobic settings, encompassing environments that exhibit extreme conditions such as elevated temperatures, acidity, salinity, or reduced oxygen concentrations. Methanogens employ various biochemical pathways, including hydrogenotrophic, acetoclastic, and methylotrophic methanogenesis, to transform substrates such as hydrogen, acetate, and methylated compounds into methane. Methanotrophic microorganisms, such as specific bacteria and archaea, utilise methane as an energy and carbon source, reducing methane emissions and impacting global methane budgets. Methanotrophs are vital in controlling methane levels in extreme environments by converting methane into carbon dioxide and their involvement in methane-driven ecological systems. Methanogens form syntrophic associations with various microorganisms, including acetogenic and sulphate-reducing bacteria, within intricate microbial consortia. Syntrophic interactions play a crucial role in the degradation of intricate organic compounds and fermentative intermediates by methanogens, thereby facilitating the process of methanogenesis in environments characterised by a scarcity of substrates (Bonfanti et al. 2018, Ciaramella et al. 2005, Dion 2023, Gast et al. 2009, Gupta et al. 2023, Kochhar et al. 2022, Nie et al. 2021, Tian et al. 2018, Zhou et al. 2024). Microbial diversity and community stability in extreme habitats are influenced by methanogens, which occupy specific ecological niches and engage in symbiotic or competitive interactions with other microorganisms. Microorganisms' metabolic activities significantly impact microbial communities' composition and function, which affects the environment's dynamics and the ability to adapt to changes. Methanogens are crucial regulators of carbon cycling, methane dynamics, and microbial interactions in extreme environments. The metabolic adaptability, capacity to flourish in challenging environments, and ecological interdependencies of these organisms play a significant role in the functioning and stability of various habitats, spanning from deep-sea hydrothermal vents to terrestrial hot springs. It is imperative to comprehend the environmental importance of methanogens in extreme environments to gain insight into ecological processes, forecast adaptations of the environment to environmental fluctuations, and formulate approaches for environmental management and preservation.

5.7 Biotechnological and Astro Biological Implications

Examining methanogens in extreme environments offers valuable insights into the ecology of microorganisms and the cycling of biogeochemical processes and carries substantial implications for biotechnology and astrobiology. This extensive investigation examines the various uses of methanogens and their enzymes and their significance in biogas generation, bioremediation, industrial operations, and the quest for extraterrestrial existence. Methanogens are essential for biogas production by breaking down organic waste materials, such as agricultural residues, municipal solid waste, and wastewater sludge, through anaerobic digestion. Methanogens play a significant role in generating renewable energy and reducing greenhouse gas emissions resulting from the decomposition of organic waste by converting organic matter into biogas rich in methane. Bioremediation strategies can utilise methanogens to facilitate the degradation of organic pollutants and contaminants in soil and water environments (Coolen et al. 2004, Gupta et al. 2019, Johnson and Aguilera 2019,

Mehta et al. 2021, Nagarajan et al. 2022, Thompson et al. 2022). Methanogens metabolise various organic compounds, such as hydrocarbons, chlorinated solvents, and aromatic pollutants, using anaerobic degradation pathways. This process helps detoxify contaminated sites and restore the health of the ecosystem. The enzymes derived from methanogens and other extremophiles possess distinctive characteristics, such as thermostability, solvent tolerance, and resilience in adverse environmental conditions. Enzymes find utility in diverse industrial applications, including but not limited to biofuel production, food processing, pharmaceutical synthesis, and bioremediation, wherein conventional enzymes may exhibit reduced efficacy. Methanogens in high-temperature environments exhibit remarkable thermostability, rendering them highly suitable for facilitating reactions under elevated temperatures. Methanogens possess thermostable enzymes that have the potential to augment the efficacy and output of industrial operations, such as starch liquefaction, cellulose hydrolysis, and bioethanol fermentation, even in challenging environmental circumstances. Methanogens that flourish in highly demanding environments on Earth can be considered potential models for life forms in extraterrestrial habitats, such as Mars, Europa, and Enceladus, where comparable conditions may exist. Examining the physiological adjustments of methanogens to harsh environments can provide valuable insights for astrobiological inquiries and direct the quest for extraterrestrial life. Methane serves as a significant biosignature gas, potentially serving as an indicator of the existence of microbial life on celestial bodies beyond Earth that possess favourable environmental circumstances. Identifying methane within the atmosphere of exoplanets or the subsurface of icy moons can yield compelling empirical support for the presence of methanogenic microorganisms or other physiological phenomena (Castillo and Villafania 2024, Chaudhary et al. 2022, Jadhav et al. 2022, Kumar et al. 2018, Lam et al. 2020, Mafra et al. 2022, Maslać et al. 2022, Passarini et al. 2022, Pati et al. 2023, Peeples 2014, Salwan and Sharma 2020, Shrestha et al. 2019). The implications of methanogens in extreme environments have extensive and complex biotechnological and astrobiological implications. Methanogens present significant prospects for advancing renewable energy technologies, environmental remediation strategies, and exploring extraterrestrial life, owing to their involvement in biogas production, bioremediation, industrial applications of extremophile enzymes, and their relevance to astrobiology. Scientists can utilise the distinctive abilities of methanogens and their enzymes to tackle urgent challenges on Earth while investigating the possibility of life in the universe.

5.8 Future Directions

Within the dynamic realm of scientific investigation, examining methanogens in highly challenging habitats persists in revealing fresh perspectives on microbial adaptation, biogeochemical cycling, and astrobiology. As we pursue enhanced comprehension, it is crucial to delineate forthcoming trajectories that drive research in novel avenues, foster interdisciplinary cooperation, and contemplate methanogens' extraordinary resilience and adaptability. The rapid progress in high-throughput sequencing technologies presents unparalleled prospects for investigating the genomic variability of methanogens in challenging ecological conditions. Subsequent investigations have the potential to provide further insight into the genomic underpinnings of adaptive characteristics, metabolic pathways, and stress response mechanisms across various methanogenic lineages. By combining omics methodologies with systems biology and ecological modelling, a comprehensive understanding of the environmental interactions, community dynamics, and ecological functions facilitated by methanogens can be achieved. Researchers can understand the unique characteristics of microbial communities in extreme habitats by analysing the intricate networks of microbial metabolism and interactions between different species. The potential for methanogenic life in extraterrestrial environments may be the subject of future research, particularly in light of the emergence of space exploration missions such as Mars rovers and missions to icy moons. The field of astrobiology will utilise a combination of different methods, such as remote sensing, geochemical analysis, and habitat simulations, to explore signs of life and evaluate the suitability of extraterrestrial planets. The investigation of methanogens and their enzymes in renewable energy generation, bioremediation, and industrial applications continues to be a subject of ongoing scholarly inquiry. Potential future endeavours could encompass the manipulation of methanogens to augment methane production, refining bioremediation methodologies, and the advancement of innovative biocatalysts for industrial utilisation. Integrating fields such as microbiology, geochemistry, ecology, bioinformaticians, and engineering is imperative to effectively tackle intricate inquiries that arise at the convergence of microbiology, ecology, and biogeochemistry. By cultivating collaborative alliances, scholars can harness a wide range of knowledge, approaches, and viewpoints to address complex issues in methanogen research. By incorporating interdisciplinary training and education programmes, a fresh cohort of scientists can be nurtured and equipped with the expertise and understanding to address research inquiries that span multiple fields. Collaborative efforts across different fields can promote originality, ingenuity, and adaptability in tackling worldwide issues, such as climate change, environmental sustainability, and astrobiology. Methanogens demonstrate remarkable resilience and adaptability to flourish in highly challenging environments worldwide. Methanogens endure and thrive in various environments, ranging from the intense heat of hydrothermal vents to the frigid expanses of polar regions. They are crucial in shaping ecological systems and significantly influence biogeochemical cycles. The pivotal role of these organisms in Earth's biosphere and beyond is underscored by their metabolic diversity, stress tolerance, and ecological significance. As we contemplate the future of scientific investigation, methanogens serve as symbols of flexibility and fascination, providing a deep understanding of the boundaries of life and the possibility of extraterrestrial environments. Through the adoption of interdisciplinary collaboration, the exploration of technological advancements, and the cultivation of a mindset of curiosity and exploration, we can decipher the enigmas surrounding methanogens and uncover the enigmas surrounding life in extreme environments. Within the vast array of life on Earth and in the universe, methanogens serve as a reminder of the extraordinary ability of microbial organisms to withstand challenges and the limitless opportunities for exploration that await us. As we explore unexplored domains, it is imperative to acknowledge and appreciate the robustness and adaptability of methanogens as guiding principles that shed light on the trajectory towards novel realms of knowledge and comprehension.

5.9 Conclusion

Within the extensive array of life forms, methanogens manifest as extraordinary microorganisms, challenging traditional conceptions of habitability by flourishing in highly challenging environments that would typically dissuade the majority of life forms. The remarkable adaptability of microbial life is demonstrated by their capacity to endure and thrive in environments marked by elevated temperatures, acidity, salinity, low oxygen levels, and other demanding circumstances. As we further investigate methanogens in extreme environments, several significant insights and implications become apparent. The resilience of methanogens exemplifies the remarkable diversity and adaptability of life on Earth. Researchers acquire valuable insights into microbial ecology, evolution, and biogeochemical cycling principles by revealing the molecular mechanisms and physiological adaptations that allow methanogens to withstand extreme conditions. Furthermore, examining methanogens in highly challenging habitats provides invaluable insights into the field of astrobiology and the quest for extraterrestrial life. By examining the boundaries of life and the possibility of microbial habitats in highly challenging environments, scholars can contribute to developing forthcoming space exploration endeavours and analysing data derived from extraterrestrial settings. In addition, methanogens are essential in biogeochemical cycles as they influence the movements of carbon, hydrogen, and other elements within terrestrial and aquatic environments. The metabolic processes of these organisms play a crucial role in generating and utilising methane. This highly potent greenhouse gas substantially impacts climate change and the global carbon cycle. In summary, examining methanogens in extreme environments sheds light on the remarkable adaptability of life. It provides significant insights into the boundaries of habitability on Earth and other celestial bodies. Scholars enhance our understanding of microbial ecology and evolution by exploring and comprehending methanogens, facilitating forthcoming advancements in astrobiology, biotechnology, and environmental science. As the exploration of microbial life progresses, methanogens serve as a testament to life's remarkable resilience, adaptability, and limitless capacity in exceedingly challenging habitats.

List of Abbreviations

1. pH - Potential of Hydrogen
2. CO_2 - Carbon Dioxide
3. O_2 - Oxygen
4. H_2 - Hydrogen
5. DNA - Deoxyribonucleic Acid
6. RNA - Ribonucleic Acid
7. ATP - Adenosine Triphosphate
8. NADH - Nicotinamide Adenine Dinucleotide

9. FADH2 - Flavin Adenine Dinucleotide
10. NASA - National Aeronautics and Space Administration
11. AMD - Acid Mine Drainage
12. pH - Potential of Hydrogen
13. NASA - National Aeronautics and Space Administration
14. H_2 - Molecular Hydrogen
15. CO_2 - Carbon Dioxide
16. CH4 - Methane
17. ATP - Adenosine Triphosphate
18. CoM - Coenzyme M
19. CoB - Coenzyme B
20. MCR - Methyl-coenzyme M Reductase

References

Alpana, Singh, Vishwakarma, P., Adhya, T. K., Inubushi, K. and Dubey, S. K. (2017). Molecular ecological perspective of methanogenic archaeal community in rice agroecosystem. Science of The Total Environment 596–597: 136–46. doi: https://doi.org/10.1016/j.scitotenv.2017.04.011.

Besse, Alison, Jean Peduzzi, Sylvie Rebuffat and Alyssa Carré-Mlouka. (2015). Antimicrobial peptides and proteins in the face of extremes: lessons from archaeocins. Biochimie 118: 344–55. doi: https://doi.org/10.1016/j.biochi.2015.06.004.

Bonfanti, Alessandra, Christophe Lecomte, Brenda J. Little and Jason S. B. T. Reference Module in Materials Science and Materials Engineering Lee. (2018). Bioactive Environments: Corrosion☆. Elsevier.

Castillo, Hugo and Mitchell Villafania. (2024). Bacterial Biodiversity. pp. 793–801 in, edited by S. M. B. T.-E. of B. Third E. Scheiner. Oxford: Academic Press.

Cena, Jéssica Alves de, Aline Belmok, Cynthia Maria Kyaw and Naile Dame-Teixeira. (2024). The archaea domain: exploring historical and contemporary perspectives with *in silico* primer coverage analysis for future research in dentistry. Archives of Oral Biology 161: 105936. doi: https://doi.org/10.1016/j.archoralbio.2024.105936.

Chastain, Brendon K. and Timothy A. Kral. (2010). Zero-valent iron on mars: an alternative energy source for methanogens. Icarus 208(1): 198–201. doi: https://doi.org/10.1016/j.icarus.2010.02.024.

Chaudhary, Srishti, Sukrampal Yadav, Ramandeep Singh, Chetan Sadhotra and Sunil A. Patil. (2022). Extremophilic electroactive microorganisms: promising biocatalysts for bioprocessing applications. Bioresource Technology 347: 126663. doi: https://doi.org/10.1016/j.biortech.2021.126663.

Ciaramella, Maria, Alessandra Napoli and Mosè Rossi. (2005). Another extreme genome: how to live at pH 0. Trends in Microbiology 13(2): 49–51. doi: https://doi.org/10.1016/j.tim.2004.12.001.

Coolen, Marco J. L., Ellen C. Hopmans, Irene, W., Rijpstra, C., Gerard Muyzer, Stefan Schouten et al. (2004). Evolution of the methane cycle in Ace Lake (Antarctica) during the Holocene: response of methanogens and methanotrophs to environmental change. Organic Geochemistry 35(10): 1151–67. doi: https://doi.org/10.1016/j.orggeochem.2004.06.009.

Dion, Patrice. (2023). Microbiology of Extreme Soil Environments. pp. 494–511 in, edited by M. J. Goss and M. B. T.-E. of S. in the E. Second E. Oliver. Oxford: Academic Press.

Dopffel, Nicole, Biwen Annie An-Stepec, Petra Bombach, Martin Wagner and Evan Passaris. (2024). Microbial life in salt caverns and their influence on H_2 storage – current knowledge and open questions. International Journal of Hydrogen Energy 58: 1478–85. doi: https://doi.org/10.1016/j.ijhydene.2024.01.334.

Fenchel, T., King, G. M. and Blackburn, T. H. (2012). Chapter 8 - Microbial Biogeochemistry and Extreme Environments. pp. 143–61 in, edited by T. Fenchel, G. M. King, and T. H. B. T.-B. B. Third E. Blackburn. Boston: Academic Press.

Gast, Rebecca J., Robert W. Sanders and David A. Caron. (2009). Ecological strategies of protists and their symbiotic relationships with prokaryotic microbes. Trends in Microbiology 17(12): 563–69. doi: https://doi.org/10.1016/j.tim.2009.09.001.

Gupta, Abhishek, Jayeeta Sarkar and Pinaki Sar. (2019). Chapter 33 - Understanding the Structure and Function of Extreme Microbiome Through Genomics: Scope and Challenges. pp. 581–610 in, edited by S. Das and H. R. B. T.-M. D. in the G. E. Dash. Academic Press.

Gupta, Rohan, Shradhanjali Sethi, Ankita Bharshankh, Rojalin Sahu and Rima Biswas. (2023). A perspective on bio-electrochemical system (BES) as a tool for boosting the performance of the nonperforming anaerobic units. Cleaner and Circular Bioeconomy 6: 100066. doi: https://doi.org/10.1016/j.clcb.2023.100066.

Holmes, D. E. and Smith, J. A. (2016). Chapter One - Biologically Produced Methane as a Renewable Energy Source. pp. 1–61 in Vol. 97, edited by S. Sariaslani and G. B. T.-A. in A. M. Michael Gadd. Academic Press.

Jabbari, Farzaneh, Babak Akbari and Lobat Tayebi. (2024). Chapter 9 - Functionalized Archaeosomes for Cancer Therapy. pp. 181–95 in, edited by H. Barabadi, E. Mostafavi, and C. B. T.-F. N. for C. R. Mustansar Hussain. Academic Press.

Jadhav, Laxmi, Vrushali Phalke, Stutee Panse, Smita Patil and Ashok Bankar. (2022). Chapter 12 - biodiversity of cold-adapted extremophiles from antarctica and their biotechnological potential. pp. 231–65 in, edited by A. Gunjal and S. B. T.-M. D. in H. Shinde. Academic Press.

Johnson, D. Barrie and Angeles Aguilera. (2019). Extremophiles and Acidic Environments. pp. 206–27 in, edited by T. M. B. T.-E. of M. Fourth E. Schmidt. Oxford: Academic Press.

Joseph, Antony. (2023). Chapter 5 - Extremophiles—Organisms That Survive and Thrive in Extreme Environmental Conditions. pp. 201–54 in, edited by A. B. T.-W. W. in the S. S. Joseph. Elsevier.

Kochhar, Nikita, Kavya I. K., Shrashti Shrivastava, Anshika Ghosh, Varunendra Singh Rawat, Kushneet Kaur Sodhi et al. (2022). Perspectives on the microorganism of extreme environments and their applications. Current Research in Microbial Sciences 3: 100134. doi: https://doi.org/10.1016/j.crmicr.2022.100134.

Koul, Bhupendra, Roni Chaudhary and Pooja Taak. (2021). Chapter 12 - extremophilic microbes and their application in bioremediation of environmental contaminants. pp. 115–28 in Microbe Mediated Remediation of Environmental Contaminants, edited by A. Kumar, V. K. Singh, P. Singh, and V. K. Mishra. Woodhead Publishing.

Kumar, Ashutosh, Anwar Alam, Deeksha Tripathi, Mamta Rani, Hafeeza Khatoon, Saurabh Pandey et al. (2018). Protein adaptations in extremophiles: an insight into extremophilic connection of mycobacterial proteome. Seminars in Cell & Developmental Biology 84: 147–57. doi: https://doi.org/10.1016/j.semcdb.2018.01.003.

Lam, Theo Y. C., Ran Mei, Zhuoying Wu, Patrick K. H. Lee, Wen-Tso Liu and Po-Heng Lee. (2020). Superior resolution characterisation of microbial diversity in anaerobic digesters using full-length 16S RRNA gene amplicon sequencing. Water Research 178: 115815. doi: https://doi.org/10.1016/j.watres.2020.115815.

Li, Yong, Zhou Zhang, Jieyu Tang, Mingxing Zhao, Zhenxing Huang, Wansheng Shi et al. (2022). The enhanced biological methanation performance in trickle-bed reactor by different continuous hydrogen supply from nano-zero valent iron corrosion. Journal of Water Process Engineering 50: 103292. doi: https://doi.org/10.1016/j.jwpe.2022.103292.

Liu, Can, Huan Li, Yuyao Zhang, Dandan Si and Qingwu Chen. (2016). Evolution of microbial community along with increasing solid concentration during high-solids anaerobic digestion of sewage sludge. Bioresource Technology 216: 87–94. doi: https://doi.org/10.1016/j.biortech.2016.05.048.

Mafra, Denise, Marcia Ribeiro, Larissa Fonseca, Bruna Regis, Ludmila F. M. F. Cardozo, Henrique Fragoso dos Santos et al. (2022). Archaea from the gut microbiota of humans: could be linked to chronic diseases? Anaerobe 77: 102629. doi: https://doi.org/10.1016/j.anaerobe.2022.102629.

Maier, Raina M. (2009). Chapter 7 - Extreme Environments. pp. 123–34 in, edited by R. M. Maier, I. L. Pepper, and C. P. B. T.-E. M. Second E. Gerba. San Diego: Academic Press.

Maier, Raina M. and Julia W. Neilson. (2015). Chapter 7 - Extreme Environments. pp. 139–53 in, edited by I. L. Pepper, C. P. Gerba, and T. J. B. T.-E. M. Third E. Gentry. San Diego: Academic Press.

Margesin, Rosa and Vanya Miteva. (2011). Diversity and ecology of psychrophilic microorganisms. Research in Microbiology 162(3): 346–61. doi: https://doi.org/10.1016/j.resmic.2010.12.004.

Maslać, Nevena, Chandni Sidhu, Hanno Teeling, Tristan Wagner and Markus W. Ribbe. (2022). Comparative transcriptomics sheds light on remodeling of gene expression during diazotrophy in the thermophilic methanogen methanothermococcus thermolithotrophicus. MBio 13(6). doi: https://doi.org/10.1128/mbio.02443-22.

Mehta, Pooja, Kalyani Deshmukh, Sumit Singh Dagar, Prashant K. Dhakephalkar and Vikram B. Lanjekar. (2021). Genome sequencing and analysis of a psychrotrophic methanogen Methanosarcina sp. Nov. MSH10X1 cultured from methane hydrate deposits of Krishna Godavari Basin of India. Marine Genomics 59: 100864. doi: https://doi.org/10.1016/j.margen.2021.100864.

Mohanty, Anee, Shilpa and Sumer Singh Meena. (2022). Chapter 4 - Microbial Adaptation to Extreme Temperatures: An Overview of Molecular Mechanisms to Industrial Application. pp. 115–39 in, edited by N. K. Arora, S. Agnihotri, and J. B. T.-E. and T. I. A. Mishra. Academic Press.

Moissl-Eichinger, Christine, Manuela Pausan, Julian Taffner, Gabriele Berg, Corinna Bang and Ruth A. Schmitz. (2018). Archaea are interactive components of complex microbiomes. Trends in Microbiology 26(1): 70–85. doi: https://doi.org/10.1016/j.tim.2017.07.004.

Nagarajan, Viji, Hsin-Chi Tsai, Jung-Sheng Chen, Bashir Hussain, Suprokash Koner, Zeng-Yei Hseu et al. (2022). Comparison of bacterial communities and their functional profiling using 16S RRNA gene sequencing between the inherent serpentine-associated sites, hyper-accumulator, downgradient agricultural farmlands, and distal non-serpentine soils. Journal of Hazardous Materials 431: 128557. doi: https://doi.org/10.1016/j.jhazmat.2022.128557.

Nie, Erqi, Pinjing He, Hua Zhang, Liping Hao, Liming Shao and Fan Lü. (2021). How does temperature regulate anaerobic digestion? Renewable and Sustainable Energy Reviews 150: 111453. doi: https://doi.org/10.1016/j.rser.2021.111453.

Passarini, Michel Rodrigo Zambrano, Alysson Wagner Fernandes Duarte, Luiz Henrique Rosa, Valéria Maia de Oliveira and Júlia Ronzella Ottoni. (2022). Chapter 10 - Extremofuels: Production of Biofuels by Extremophile Microbes as an Alternative to Avoid Climate Change Effects. pp. 237–56 in, edited by A. Kumar, J. Singh, and L. F. R. B. T.-M. U. C. C. Ferreira. Woodhead Publishing.

Pati, Swatilekha, Somok Banerjee, Aveepsa Sengupta, Jayshree Sarma, Shakila Shaheen, Shivendra Tenguria et al. (2023). Chapter 16 - Adaptation Strategies of Thermophilic Microbes. pp. 231–49 in Developments in Applied Microbiology and Biotechnology, edited by A. Kumar and S. B. T.-B. S. in the H. E. Tenguria. Academic Press.

Peeples, Tonya L. (2014). 10 - Bioremediation Using Extremophiles. pp. 251–68 in, edited by S. B. T.-M. B. and B. Das. Oxford: Elsevier.

Rogers, Rudy. (2015). Chapter Ten - Martian Hydrate Feasibility: Extending Extreme Seafloor Environments. pp. 323–60 in, edited by R. B. T.-O. G. H. Rogers. Boston: Gulf Professional Publishing.

Salwan, Richa and Vivek Sharma. (2020). Chapter 2 - Physiology of Extremophiles. pp. 13–22 in, edited by R. Salwan and V. B. T.-P. and B. A. of E. Sharma. Academic Press.

Santos-Pereira, Cátia, Joana Sousa, Sara C. Silvério, Marta Filipa Simões, André Antunes and Lígia R. Rodrigues. (2024). Chapter 4 - Metagenomics to unravel the microbial biodiversity and biotechnological potential of extreme high salinity environments. pp. 77–130 in Developments in Applied Microbiology and Biotechnology, edited by M. P. B. T.-F. M. Shah. Academic Press.

Shao, Nana and William, B. B. T. Reference Module in Life Sciences Whitman. (2023). Archaeal Genetics. Elsevier.

Shrestha, Namita, Govinda Chilkoor, Saurabh Sudha Dhiman, Rajesh K. Sani and Venkataramana Gadhamshetty. (2019). Chapter 2.6 - Extremophile biology for microbial electrochemistry applications. pp. 353–74 in Biomass, Biofuels and Biochemicals, edited by S. V. Mohan, S. Varjani, and A. B. T.-M. E. T. Pandey. Elsevier.

Singh, Rachana and Paras Porwal. (2022). Chapter 3 - Innovative Technologies for Enzyme Production from Extremophilic Microbes. pp. 31–38 in, edited by M. B. T.-M. E. Kuddus. Academic Press.

Sinha, Navita, Sudip Nepal, Timothy Kral and Pradeep Kumar. (2017). Survivability and growth kinetics of methanogenic archaea at various pHs and pressures: implications for deep subsurface life on mars. Planetary and Space Science 136: 15–24. doi: https://doi.org/10.1016/j.pss.2016.11.012.

Thompson, Thomas P., Julianne Megaw, Stephen A. Kelly, Jason Hopps and Brendan F. Gilmore. (2022). Chapter One - Microbial Communities of Halite Deposits and Other Hypersaline Environments. pp. 1–32 in Vol. 120, edited by G. M. Gadd and S. B. T.-A. in A. M. Sariaslani. Academic Press.

Tian, Hailin, Ioannis A. Fotidis, Enrico Mancini, Laura Treu, Ahmed Mahdy, Mercedes Ballesteros et al. (2018). Acclimation to extremely high ammonia levels in continuous biomethanation process and the associated microbial community dynamics. Bioresource Technology 247: 616–23. doi: https://doi.org/10.1016/j.biortech.2017.09.148.

Welte, Cornelia and Uwe Deppenmeier. (2014). Bioenergetics and anaerobic respiratory chains of aceticlastic methanogens. Biochimica et Biophysica Acta (BBA) - Bioenergetics 1837(7): 1130–47. doi: https://doi.org/10.1016/j.bbabio.2013.12.002.

Zhou, Pan, Dunjie Li, Cong Zhang, Qian Ping, Lin Wang and Yongmei Li. (2024). Comparison of different sewage sludge pretreatment technologies for improving sludge solubilization and anaerobic digestion efficiency: a comprehensive review. Science of The Total Environment 921: 171175. doi: https://doi.org/10.1016/j.scitotenv.2024.171175.

Chapter 6

Anaerobic Bioenergy Production

Harnessing Biomethane Potential

Harshika Ghadge, Ayushi Shah, Abhishant Nigam and *Shadab Ahmed**

6.1 Introduction

Fossil fuels powered the industrial revolution, which had a tremendous global impact on socioeconomic and technological progress (Janaun and Ellis 2010). However, the increased reliance on non renewable energy sources has had major environmental effects, including climate change, pollution, and greenhouse gas emissions (GHG). To counteract this effect, people are shifting their focus to sustainable energy strategies that prioritize the utilization of renewable resources like geothermal, hydro, solar, biomass, and wind energy. Among them, biomass has emerged as a major resource (Röder and Welfle 2018). Anaerobic Digestion (AD) converts organic wastes into biomass energy, offering a promising alternative to fossil fuels (Mourshed et al. 2023).

Anaerobic digestion uses bacteria to decompose organic waste in the absence of oxygen, producing biogas, which is a mixture of methane (CH_4) and carbon dioxide. The AD process contributes to waste management while reducing its environmental impact (Van Lier et al. 2008).

Hydrolysis, acidogenesis, acetogenesis, and methanogenesis are the processes that transform organic substances into simpler molecules, ultimately producing methane and CO_2. Hydrolysis converts complex organic polymers like carbohydrates, proteins, and lipids into easily soluble molecules (Park et al. 2005).

Department of Biotechnology, Savitribai Phule Pune University, Pune.
Emails: harshikaghadge@gmail.com, ayushicshah02@gmail.com, abhishantnigam10@gmail.com
* Corresponding author: sahmed@unipune.ac.in

DOI: 10.1201/9781003506898-6

During acidogenesis, these soluble compounds are converted to fatty acids (VFAs), alcohol, hydrogen (H_2), and carbon dioxide (CO_2). Acetogenesis then converts VFA and alcohol into acetic acid, H_2 and CO_2, and finally, methanogenesis converts these intermediates into methane. Solid waste, industrial waste and biomass are considered best substrates for anaerobic digestion (Arelli et al. 2022). Many factors, such as the composition and properties of the feedstock, the total composition of the waste, Hydraulic Retention Time (HRT), pH level, temperature and organic loading rate affects the efficiency of biogas production. By optimizing these parameters, the production rate and yield can be increased (Hashemi et al. 2021). The purification process involves removing impurities such as hydrogen sulphide (H_2S), ammonia (NH_3) and siloxanes, which can affect the efficiency and safety of biogas technology (Vu et al. 2022).

The biomethane industry, although currently small with a global production of 3.5 million tons of oil equivalent, is receiving significant attention for its potential to provide clean energy through existing processes. Most production occurs in Europe and North America, where countries such as Denmark and Sweden account for a large share of biogas/biomethane in all-natural gas production. Other regions such as Brazil, China and India have the potential to develop rapidly (IEA (2020), Outlook for biogas and biomethane: Prospects for organic growth, IEA, Paris).

Government initiatives in Germany, Italy, the Netherlands, the United Kingdom, and Brazil promote its use, particularly in the transportation industry (2016 European Biogas Association Biomethane in Transport). Carbon dioxide is the primary byproduct of regeneration, which can be reused in industry or agriculture or mixed with hydrogen to make further methane. Biomethane generation via thermal gasification of biomass has not yet been achieved. There is enormous room for innovation and cost savings.

6.2 Pretreatment Methods

Plant derived lignocellulosic biomass consists of cellulose, hemicellulose, and lignin (Balat 2011). Agricultural residues (e.g., maize stover, wheat straw), forest residues (e.g., wood chips, sawdust), energy crops (e.g., switchgrass, miscanthus), and municipal domestic waste, containing vegetable waste, paper, cardboard, etc., are major sources for lignocellulosic biomass (Sun and Cheng 2002). These abundantly available lignocellulosic biomass is a promising feedstock for biomethane production and can reduce greenhouse gas emissions.

Using anaerobic digestion, microbes degrade the complex carbohydrates in lignocellulosic biomass, producing methane-rich biogas. Anaerobic digestion (AD) can degrade a wide range of feedstocks, such as crop and livestock wastes, food waste, and lignocellulosic materials (Hagos et al. 2016). But, the amount of methane produced varies significantly depending on the type of substrate used (Anacleto et al. 2024). For example, sugar and starch crops can yield high methane quantities approximately reaching up to 450 mL of CH_4 per gram of solids, whereas lignocellulosic biomass yields 330 mL or less CH_4 per gram of volatile solids (Frigon et al. 2011). This is due to the recalcitrant nature of lignocellulose. Hence demands for pretreatment to increase the digestibility which will increase biogas production (Rahimi et al. 2018) (Fig. 6.1).

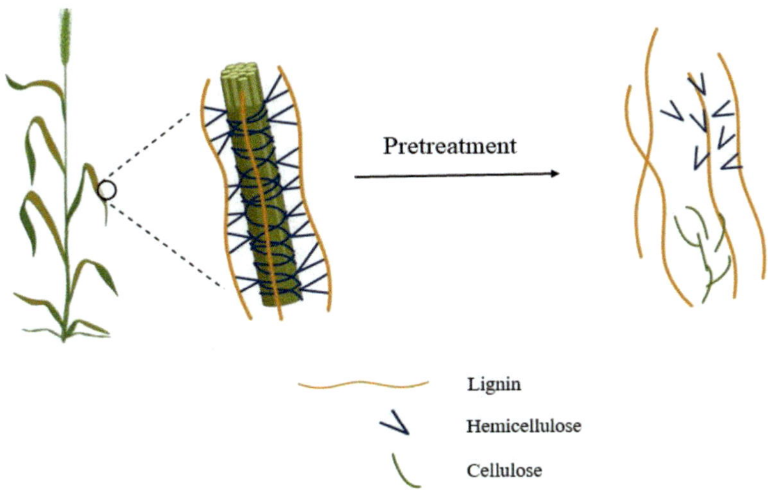

Figure 6.1. Effect of pretreatment on lignocellulosic biomass.

Lignocellulosic biomass is composed of cellulose, hemicellulose, and lignin, which form a complex and recalcitrant structure. Other than the structure, lignocellulosic biomass may produce inhibitory compounds such as phenolics, furans and organic acids (Zha et al. 2014), during anaerobic digestion. Various pretreatment methods like physical, chemical, or biological treatments are adopted to break down lignin and hemicellulose, exposing cellulose for enzymatic hydrolysis.

6.2.1 Physical Method

Physical methods involve mechanical treatments such as milling, grinding, and irradiation for size reduction and to increase the surface area and accessibility of biomass to enzymes or microorganisms (Kumar et al. 2009). Mechanical methods reduce crystallinity. It also affects the water retention capacity of biomass. Although the quality of methane produced is not affected as much the increase in the methane produced is significant (Fernandez et al. 2020). These methods improve water retention and enhance digestibility. Physical methods are relatively simple and cost-effective. However, physical methods demand high energy consumption and monitor equipment wear and maintenance (Zhang and El-Mashad 2012).

6.2.2 Chemical Methods

Chemical methods comprise acid pretreatment, alkali pretreatment, oxidation pretreatment, Ionic Liquid (IL) pretreatment, and organosolv pretreatment (Mendes Carlos et al. 2015).

Acid pretreatment involves the use of both inorganic acids and organic acids to separate lignin and hydrolyze cellulose and hemicellulose. It converts the maximum available substrate, reaching over 90% conversion rate for both cellulose and hemicellulose (Sai Bharadwaj et al. 2023).

Concentrated acid pretreatment poses some challenges as it is relatively slow, has high operational costs, and equipment corrosion (Brodeur et al. 2011). Due to this it is not widely accepted commercially. To reduce these risks, dilute acid pretreatment is adopted, however, to get the desired results, dilute acid treatment should be performed under high pressure and temperature conditions. Despite drawbacks, dilute sulphuric acid pretreatment has been extensively studied and shown promising results, particularly in increasing xylose yield. Thus, to enhance its industrial viability acid recovery method should be optimized (Saha et al. 2005).

Alkali pretreatment uses NaOH, KOH, and $Ca(OH)_2$ to effectively dissolve lignin and increase internal surface area, which improves enzymatic hydrolysis. It provides relatively gentle reaction conditions with negligible sugar degradation (Donkor et al. 2022). Alkalis are difficult to recover and recycle, however, $Ca(OH)_2$ can be easily recovered compared to others. Optimizing pretreatment conditions and alkali recovery procedures may improve industrial applications (Chen et al. 2017).

6.2.3 Physicochemical Methods

6.2.3.1 Steam Explosion Pretreatment

Steam explosion is a widely used physical and chemical pretreatment method that briefly treats biomass with high-temperature steam and then quickly depressurizes. This process destroys the fibrous structure of biomass, increases its digestibility by destroying the lignocellulosic matrix, reduces the crystallinity of cellulose and promotes the removal of hemicellulose (Pang et al. 2013). The main benefits of digestion include increased efficiency of enzymatic hydrolysis and increased methane production during anaerobic digestion (Bruni et al. 2010). But it also produces anti-inflammatory chemicals such as furans and phenols, and these need to be controlled to prevent negative effects on the digestive system (Pavičić et al. 2022).

6.2.3.2 Ammonia Fibre Explosion (AFEX)

In AFEX, biomass is treated at high pressure and temperature with liquid ammonia, and then the pressure quickly released (Chundawat et al. 2020). Instant decrease in pressure and temperature leads to increase in surface area of cellulose and decreases its crystallinity, thus making it more accessible to enzymes. AFEX is especially good at eliminating lignin and hemicellulose, which increases the amount of methane that can be produced from herbaceous plants and agricultural waste. This process recovers ammonia and does not use antibiotics which makes it affordable and environmentally friendly (Galbe and Wallberg 2019).

6.2.3.3 Carbon Dioxide (CO_2) Explosion Pretreatment

A CO_2 explosion is similar to a gas explosion, but uses carbon dioxide under pressure, producing carbonic acid. This method increases the hydrolysis activity of hemicellulose and destroys the lignocellulosic structure without causing chemical reactions (Toscan et al. 2017). It is cheaper and less toxic than AFEX, making it a suitable choice for large-scale applications (Brodeur et al. 2011). The main impact of CO_2 cracking is the importance of the rate of enzymatic hydrolysis resulting in more methane (Galbe and Wallberg 2019).

6.2.3.4 Sulphur Dioxide (SO_2) Explosion Pretreatment

SO_2 explosion uses sulphur dioxide as an acid catalyst to dissolve hemicellulose and partially hydrolyze cellulose (Mackie et al. 1985). This method works well at temperatures below breaking point and especially on softwood materials. However, due to the corrosive nature of sulphur dioxide, the process requires efficient equipment and can produce many degrading chemicals that must be controlled (Wang et al. 2018). The main advantage of SO_2 pretreatment is the efficient conversion of cellulose into fermentable sugars, thereby increasing methane production.

6.2.3.5 Electrocatalytic Pretreatment

Electrocatalysis uses an electric field to start a catalytic reaction for degrading lignocellulosic biomass. This approach is ecologically acceptable and prevents secondary pollution (Amin et al. 2017). The procedure encourages cellulose hydrolysis by using electrodes which consist of substances like SnO_2 and CeO_2. The primary result of electrocatalysis is an increase in lignin and hemicellulose degradation, which increases the production of methane during anaerobic digestion (Baruah et al. 2018). By increasing access to cellulase hydrolysis, the pretreatment procedure influences the synthesis of biomethane and enhances the overall outcomes of anaerobic digestion. Methane production may be increased, reduction in digestion time and process safety can be raised with better purification (Yang et al. 2024) (Table 6.1).

Table 6.1. Overview of the pretreatment methods and its advantages and disadvantages.

Pretreatment Method	Principle	Advantages	Disadvantages
Physical Pretreatment			
Mechanical	Reduces particle size using milling or grinding.	Increases surface area, facilitates further processing.	High energy consumption.
Steam Explosion	Biomass is treated with high-pressure steam and then rapidly decompressed	Disrupts lignin structure, enhances enzyme accessibility.	Can lead to the formation of inhibitors like furfural.
Chemical Pretreatment			
Acid Hydrolysis	Acids are used to hydrolyze hemicellulose.	Effective hemicellulose breakdown.	Highly corrosive and hazardous
Alkaline Hydrolysis	Bases are employed to solubilize lignin and hemicellulose.	Enhance enzyme accessibility.	Long reaction time and the need to recover chemicals.
Physicochemical Pretreatment			
Ammonia Fibre Expansion (AFEX)	Liquid ammonia treatment at moderate temperature and pressure.	Inhibitors like furfurals are not formed.	Ammonia recovery is necessary.

6.3 Production of Biomethane by Anaerobic Digestion

Bio-methanation is a highly efficient process that involves the conversion of organic waste into valuable biomethane and other byproducts through the utilization of microorganism enzymes which act on a substrate–often a waste material (Liebetrau et al. 2017) such as lignocellulosic biomass, waste water, sewage sludge, etc. The mean biomethane yields for various types of sludge on a global scale exhibit distinct value. Specifically, primary sludge yields an average of 0.425 Nm^3 kg VS^{-1}, indicating a relatively higher conversion rate to biomethane. In contrast, mixed sludge demonstrates a mean yield of 0.296 Nm^3 kg VS^{-1}, positioning it as an intermediary option in terms of biomethane production potential. Waste activated sludge, with a mean biomethane yield of 0.176 Nm^3 kg VS^{-1}, presents the lowest conversion rate among the sludge types considered. These figures highlight the varying degrees of methane generation potential across different sludge categories, thereby necessitating tailored approaches for maximizing biomethane production based on the specific characteristics of each sludge variant (Calabrò et al. 2024).

Being a prominent waste-to-energy solution, biogas generation has gained substantial traction across various industries to address the increasing energy requirements. The global landscape witnesses the generation of organic matter from diverse sources such as sewer sludge, animal waste, poultry waste, agricultural residues, and municipal solid waste, among others. This organic waste, which would otherwise be considered a burden, can be repurposed to produce biomethane by undergoing a natural biological process under anaerobic conditions known as Anaerobic Digestion (AD).

Anaerobic digestion is a biochemical process that breaks down organic materials in the absence of oxygen, resulting in the production of biogas and digestate. This process is widely used for waste management and renewable energy production. The biogas, primarily composed of methane (CH_4) and carbon dioxide (CO_2), can be utilized as a sustainable energy source, while the digestate can be used as a fertilizer. The AD process is influenced by various factors including substrate composition and pre-treatment choice, temperature, pH, retention time, and microbial community dynamics.

Previous experimental studies in the field have shown a significant variation in the anaerobic digestibility of non-pretreated bio sludge sourced from pulp and paper mills. These studies have highlighted that the removal rates of Volatile Solids (VS) typically fall within the range of 21 to 55%, accompanied by specific methane yields that span from 40 to 200 mL g^{-1} VS fed. While it has been observed that pretreatment methods can enhance the digestibility of bio sludge to a certain degree, the data indicates that, in nearly all documented instances, the specific methane yield of pretreated bio sludge has not surpassed the threshold of 200 mL g^{-1} VS fed. The improvements in specific methane yield post-pretreatment generally vary between 0 and 90% when compared to non-pretreated bio sludge. Nevertheless, it is noteworthy that more substantial enhancements have typically been achieved when dealing with bio sludge that presents higher levels of complexity in terms of digestibility (Meyer and Edwards 2014).

The effectiveness of biogas production is highly influenced by the selection of feedstock and the type of substrate used in the process (Mao et al. 2017). Various trace elements and macronutrients play a crucial role in ensuring the stability and efficiency of the anaerobic digestion process. These elements serve as essential components of enzymes and cofactors involved in the breakdown of organic matter and subsequent biogas production. Studies have proven that the addition of these key nutrients to anaerobic digesters significantly enhances the overall performance and sustainability of biogas production by supporting the growth and activity of microorganisms responsible for the decomposition of organic materials.

Microorganisms play a pivotal role in the anaerobic digestion process, where organic matter is broken down to produce biogas, a valuable energy source utilized for heat and electricity generation. The composition of the biogas is influenced by the feedstock and substrate types used in the process. The availability of macronutrients and trace elements is crucial for maintaining the stability and efficiency of biogas production (Thanh et al. 2016). Various strategies, such as co-digestion and pre-treatment methods, are employed to optimize the biogas yield (Abouelenien et al. 2014). The concept of biorefinery not only offers significant environmental benefits but also drives technological advancements leading to the creation of new products, employment opportunities, and the recycling of by-products as biofertilizers. Anaerobic digestion is now widely recognized as a cost-effective, energy-recovery, and environmentally friendly method with the potential to contribute significantly to sustainable waste management and renewable energy production (Zou et al. 2016).

Organic polymers undergo a complex breakdown process at the initiation of anaerobic digestion, yielding fundamental components such as sugars, fatty acids, and amino acids, sequentially. These monomers are subsequently metabolized by fermentative bacteria, predominantly acidogens, to generate a mixture of short-chain volatile fatty acids. Following this, acetogenic bacteria, also referred to as acetogens, convert the volatile fatty acids into acetate, carbon dioxide, and hydrogen—key organic substrates utilized by methanogenesis in the production of biomethane. The anaerobic digestion (AD) mechanism theoretically enables the decomposition of the organic constituents from various feedstock sources, including lignocellulosic materials, agricultural byproducts, and food waste, to generate biomethane.

The initial phase of biomethane production encompasses feedstock preparation, necessitating the careful selection and conditioning of organic waste materials like food scraps, farm residues, and livestock manure to facilitate microbial digestion. Subsequently, the anaerobic digestion process ensues, wherein microorganisms break down the organic matter to generate biogas, primarily composed of methane and carbon dioxide. The final procedural stage involves gas upgrading, a critical step focused on eliminating impurities such as carbon dioxide and hydrogen sulphide to refine the biogas into biomethane. This purification process is indispensable for ensuring that the biomethane meets the stringent quality standards essential for its injection into the natural gas distribution network or utilization as a viable transportation fuel option (Fig. 6.2).

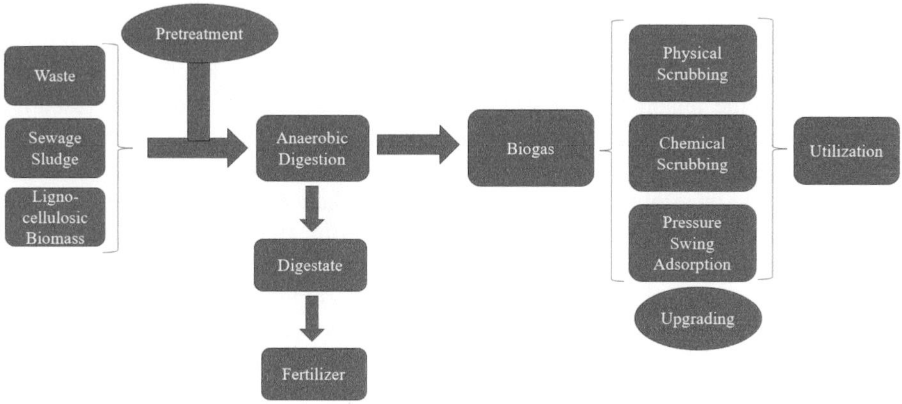

Figure 6.2. Schematic overview of biomethane production (Lozanovski et al. 2014).

After the anaerobic digestion (AD) process is complete, biogas is produced as a byproduct, containing a mixture of methane (CH_4), carbon dioxide (CO_2), and various other gases in trace amounts. This biogas is subjected to purification processes to upgrade it into biomethane. These purification techniques include methods such as pressure swing adsorption, water scrubbing, and membrane separation (Nguyen et al. 2021). Through these processes, the biogas is refined to remove impurities and increase the methane content, resulting in purified biomethane. Biomethane, once purified, becomes a valuable renewable energy source that can be utilized as a sustainable fuel alternative across various applications, contributing to a more environmentally friendly energy landscape.

The process of anaerobic digestion (AD) consists of four crucial stages: hydrolysis, acidogenesis, acetogenesis, and methanogenesis. Although methanogenesis marks the final stage of AD, it is important to note that both hydrolysis and methanogenesis are considered the rate-limiting processes within this bioconversion process (Tsigkou et al. 2022). Throughout the AD process, there is a significant reliance on a diverse array of microorganisms that play distinct roles at different stages. These anaerobic microorganisms collaborate within a complex ecosystem to effectively decompose organic materials and generate methane gas as a byproduct. A pivotal aspect of the process is the successive utilization of the end products from one stage as substrates for the subsequent phase, ensuring a seamless flow of bioconversion. For optimal efficiency in converting waste into biogas, each stage necessitates precise conditions and specific microbial communities to carry out the intricate biochemical transformations successfully. The efficiency of the AD process depends on the balance and activity of these microorganisms throughout the stages. Without these specific conditions and microbial communities, the breakdown of organic matter and the production of methane gas would not be possible (Lohani and Havukainen 2018) (Fig. 6.3).

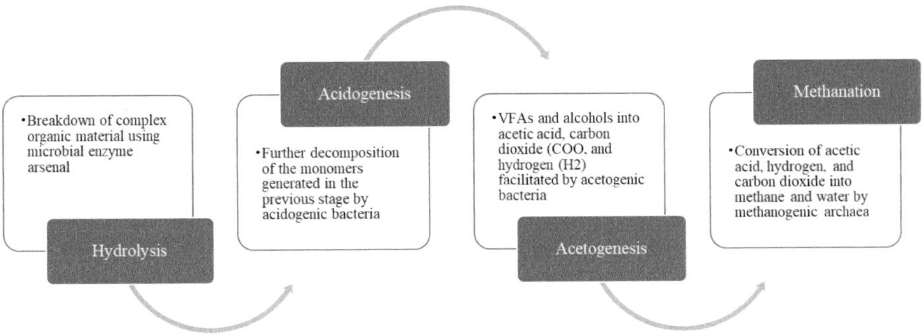

Figure 6.3. Stages of biomethane production by anaerobic digestion.

6.3.1 Hydrolysis

At the onset of anaerobic digestion, hydrolysis initiates the breakdown of intricate organic compounds such as carbohydrates, proteins, and lipids into simpler units like sugars, amino acids, and fatty acids through the action of hydrolytic enzymes. The efficiency of this initial stage is critical as it dictates the availability and accessibility of substrates for subsequent phases of the process. Mainly two anaerobic bacterial species are used for this stage:

Clostridium species: These bacteria produce extracellular enzymes that break down complex polymers such as carbohydrates, proteins, and lipids into simpler compounds like sugars, amino acids, and fatty acids.

Bacteroides species: These bacteria are known for their role in breaking down complex carbohydrates into simple sugars.

6.3.2 Acidogenesis

Subsequent to hydrolysis, acidogenesis involves the further decomposition of the monomers generated in the previous stage by acidogenic bacteria, resulting in the production of Volatile Fatty Acids (VFAs), alcohols, hydrogen (H_2), and carbon dioxide (CO_2). This stage is marked by the rapid generation of acids and other intermediary products crucial for the progression of anaerobic digestion. The microbial communities which can be used for this stage are:

Enterobacter: These bacteria ferment the products of hydrolysis into volatile fatty acids, alcohols, hydrogen, and carbon dioxide.

Lactobacillus: Involved in the production of lactic acid from the fermentation of sugars.

6.3.3 Acetogenesis

Acetogenesis marks the conversion of VFAs and alcohols into acetic acid, carbon dioxide (CO_2), and hydrogen (H_2) facilitated by acetogenic bacteria. This pivotal stage generates the necessary substrates required by methanogens to facilitate the final phase of anaerobic digestion.

Syntrophomonas: These bacteria perform syntrophic acetate oxidation, producing acetate, hydrogen, and carbon dioxide from fatty acids and alcohols.

Syntrophobacter: Involved in the conversion of propionate to acetate, hydrogen, and carbon dioxide in cooperation with hydrogenotrophic methanogens.

6.3.4 Methanogenesis

The culminating stage, methanogenesis, involves the conversion of acetic acid, hydrogen, and carbon dioxide into methane and water by methanogenic archaea. This phase is highly sensitive to environmental variables and plays a pivotal role in the production of biogas, emphasizing its significance in the overall anaerobic digestion process. The following species can be utilized for this step-

Methanosarcina spp. (Archaea): Capable of producing methane from acetate, hydrogen, and carbon dioxide.

Methanobacterium spp. (Archaea): Produce methane from hydrogen and carbon dioxide.

6.4 Factors Affecting the Biogas Production Yield, Rate and Upscaling

Factors influencing the yield, rate, and scaling of biogas production are multifaceted, with a primary challenge stemming from the intricate structure of biomass. Lignocellulosic biomass presents a significant hurdle due to its high resistance to anaerobic degradation, resulting in a suboptimal biomethane yield. The concept of "biomass recalcitrance" encapsulates the robust anti-degradation characteristics inherent in native lignocellulose, impeding the hydrolysis process in the initial phases of anaerobic digestion. This obstacle ultimately constrains the widespread commercialization of biomethane derived from lignocellulosic sources. Addressing this issue is paramount in enhancing the efficiency and viability of biogas production on a larger scale (Laiq Ur Rehman et al. 2019).

The pH range of the digester is a critical factor that significantly impacts the efficient growth of potential microorganisms in the Acid Decomposition (AD) process (Kouzi et al. 2020). The optimal pH range varies depending on factors such as substrate type, Organic Loading Rate (OLR), and digester type. Deviations from the ideal pH levels can lead to detrimental effects on the process. High pH levels can result in free ammonia toxicity, while low pH levels may lead to the formation of non-ionized sulphide and hydrogen sulphide gas, both of which can disrupt the microbial activity essential for the breakdown of organic matter.

The stability of the anaerobic digestion system is often gauged by monitoring the effluent of the reactors. Fluctuations in pH levels within the system are indicative of a lack of buffering capacity, which can trigger imbalances and hinder the overall efficiency of the process (Fig. 6.4).

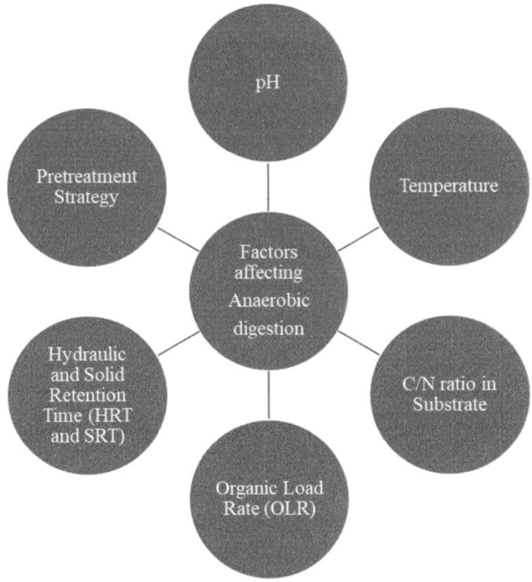

Figure 6.4. Factors affecting anaerobic digestion process.

Anaerobic digestion processes are commonly classified into psychrophilic, mesophilic, and thermophilic based on the operating temperature ranges of 10–20°C, 20–45°C, and 50–65°C, respectively (Hupfauf et al. 2018, Liu et al. 2018a, Liu et al. 2018b). Temperature plays a crucial role in substrate utilization, microbial growth, and biogas production efficiency within the digester. Fluctuations in temperature can have detrimental effects on the process, leading to the accumulation of Volatile Fatty Acids (VFAs), reduced waste volume degradation rates, and potential inactivation of crucial microorganisms essential for the breakdown of organic matter.

Psychrophilic microorganisms exhibit low metabolic rates, making them less efficient in biogas production compared to mesophilic and thermophilic counterparts. On the other hand, thermophilic microorganisms operate at high metabolic rates, accelerating the degradation of organic matter and biogas production. Mesophilic conditions offer a balance between the extremes of psychrophilic and thermophilic conditions, providing more stability, reduced sensitivity to environmental fluctuations, and requiring lower energy and cost inputs for maintaining the anaerobic digestion process.

The Carbon-to-Nitrogen (C/N) ratio plays a critical role in the anaerobic digestion (AD) process, serving as a key indicator of the relationship between carbon and nitrogen content present in the feedstock. Carbon acts as the primary energy source, facilitating the microbial breakdown of organic matter, while nitrogen is essential for supporting microbial proliferation within the AD system. For optimal microbial activity, a C/N ratio ranging between 20 to 30 is typically favoured. Deviations from this range can lead to significant consequences; lower-than-optimal ratios can elevate pH levels and introduce toxicity concerns, while higher ratios may provoke system instability and hinder the production of methane gas through methanogenesis (Braz et al. 2018).

Achieving and maintaining the appropriate C/N ratio in the AD process is paramount and can be attained through the judicious selection and digestion of varied substrates (Dong et al. 2015). Research indicates that adjustments in the C/N ratio have a direct impact on the levels of Fatty Acids (FAs) produced, with higher ratios being associated with reduced FA values. However, the efficient conversion of lignocellulosic biomass into biogas faces challenges due to slow microbial degradation rates and the presence of inhibitory compounds. To address these limitations, pre-treatment methods are employed to enhance the breakdown of complex organic compounds, increase the surface area available for enzymatic action, and improve the overall biodegradability of the biomass.

These pre-treatment techniques are generally classified into physical, chemical, thermal, and biological methods, each offering unique approaches to optimize the conversion of lignocellulosic biomass into valuable biogas. Physical methods involve the mechanical disruption of biomass structures to enhance accessibility for microorganisms, whereas chemical treatments aim to modify biomass composition and remove inhibitory substances. Thermal treatments utilize heat to break down complex molecules, while biological methods rely on the action of microorganisms to facilitate biomass degradation and gas production. By carefully selecting and implementing these pre-treatment strategies, the efficiency and effectiveness of biogas production from lignocellulosic biomass can be significantly enhanced in AD systems.

Hydraulic Retention Time (HRT) and Solid Retention Time (SRT) are fundamental factors that significantly influence the efficiency of anaerobic digestion processes. HRT is defined as the average duration that feedstock remains within the digester before being discharged, while SRT refers to the average period during which bacteria reside in the digester environment. Maintaining optimal HRT and SRT levels is crucial as deviations can lead to undesirable outcomes. A decrease in both HRT and SRT can result in the overproduction of volatile fatty acids (VFA), a reduction in alkalinity levels, and a decrease in biogas production efficiency. Conversely, higher HRT or SRT values contribute to increased methane production, effective removal of pollutants, and reduced risk of acidification; however, this comes with the trade-off of heightened operational costs (Laiq Ur Rehman et al. 2019).

The recommended range for SRT typically falls between 10 and 40 days, ensuring the balanced presence of bacteria for efficient digestion processes. On the other hand, HRT varies depending on the operating conditions, with mesophilic conditions typically ranging from 10 to 40 days and thermophilic conditions hovering around 14 days (Nouhou Moussa et al. 2023). Organic Loading Rates (OLR) play a significant role in sustaining a harmonious balance between acidogenesis and methanogenesis within the digester. It represents the quantity of organic matter loaded into the continuous digester per cubic metre daily. Excessive OLR levels can lead to the accumulation of inhibitory compounds, acidification of the medium, inhibition of methanogenic bacteria, and subsequently, a reduction in biogas yield. Therefore, maintaining a balanced and appropriate feedstock amount is imperative for achieving optimal outcomes in anaerobic digestion processes (Musa et al. 2018).

Anaerobic digestion (AD) involves harnessing anaerobic digesters to facilitate the production of biogas through the decomposition of organic matter. The success of this process heavily relies on maintaining specific temperature and pH conditions, which are conducive to the survival and activity of essential microorganisms. AD reactors can be categorized into single-stage, two-stage, or three-stage configurations, operating under either batch or continuous flow systems. While a single-stage reactor encompasses all digestion phases within a single unit, multi-stage reactors offer enhanced efficiency by segregating different stages, thus providing opportunities for optimization. Various reactor designs such as the anaerobic contact process, agitated granular sludge bed reactor, carrier-induced granular sludge bed reactor, up-flow anaerobic sludge blanket reactor, solid-state digester, and temperature-phased anaerobic digestion reactor cater to diverse needs and requirements. Selecting and maintaining suitable operating conditions are pivotal for ensuring the maximum biogas yield through efficient anaerobic digestion processes.

6.5 Applications of Biomethane

As described in the previous text, biomethane is a versatile and renewable source of energy derived from organic matter. Its sustainable property makes it a sought after, clean alternative for fossil fuels, for various applications discussed below. Biomethane is considered to be an upgraded form of biogas, having more than 97% of methane content (Herbes 2017).

6.5.1 Transportation Sector

Biomethane has emerged to be a suitable alternative fuel for transportation, especially in heavy duty vehicles like trucks, etc., with notably reduced harmful emissions (Pääkkönen et al. 2019). Biomethane can be pressurized or liquefied and can be transported using the existing systems where natural gas is utilized (Pääkkönen et al. 2019). Biomethane as transportation fuel can achieve emission reductions of up to 90% in CO_2 and 95% in particulate matter.

Apart from using biomethane as a direct source of fuel, it can also be used to produce multiple biofuels such as hydrogen, compressed biogas, dimethyl ether, liquid biogas (LBG), etc., as described in the next section. All these biofuels have a high value in the transport sector (Dahlgren 2022). Biomethane is liquefied to make LBG, or gasification is performed to produce syngas, that is further used in the production of hydrogen, DME, methanol, etc. (Dahlgren 2022). Biomethane that is up to a certain country's natural gas standard can be used in vehicles that currently use CNG (Koonaphapdeelert et al. 2020).

In summary, the benefits of using biomethane in transportation include its lower pollution levels, its cost-effectiveness compared to gasoline, diesel, and LPG, its safety due to being lighter than air—allowing it to rise and dissipate if leaked—and its low flammability (Koonaphapdeelert et al. 2020).

However, there are still many flaws to work out before full-fledged replacement of natural gas by biomethane. Other methods used to convert biomethane to methanol include- dry reforming, partial oxidation, autothermal reforming (ATR) and the Fischer-Tropsch (FT) process. Synthesis gas or syn gas is the main product of biomethane reforming process.

6.5.2 Dry Reforming

Dry reforming is a chemical process of formation of syngas. Methane reacts with carbon dioxide, as shown in the reaction below, to produce synthetic gas which is a mixture of hydrogen and carbon monoxide.

$$CH_4 + CO_2 \rightarrow 2H_2 + 2CO$$

6.5.3 Partial Oxidation Reforming

In partial oxidation reforming is used to produce hydrogen using less energy as it is moderately exothermic. H_2 and CO are formed by partial oxidation at atmospheric pressure and temperatures between 700 to 900°C. The hydrocarbons react with a limited amount of oxygen to produce syngas.

$$CH_4 + 0.5O_2 \rightarrow CO + 2H_2$$

6.5.4 Autothermal Reforming (ATR)

ATR uses both steam reforming and partial oxidation in the presence of CO_2. Methane reacts with limited oxygen as well as steam to produce syngas (Jeremiah et al. 2023). Partial oxidation is an exothermic reaction and the heat produced by it drives the reaction of methane with steam. Such a process yields higher hydrogen and consumes less oxygen (Alves et al. 2013).

6.5.5 Fischer-Tropsch (FT) Process

The Fischer-Tropsch (FT) process converts syngas into liquid hydrocarbons, i.e., fuels and lubricants. This process can be used at the industrial scale (Herz et al. 2017). Syngas is first produced from various feedstocks like coal or biomass by one of the methods described above (partial oxidation, steam reforming, etc.) and the syngas then, in the presence of a catalyst in a reactor, gets converted into liquid hydrocarbons (Chai et al. 2023). The resulting hydrocarbons are separated and refined into various final products like waxes, gasoline, lubricants, etc.

6.6 Production of Other Biofuels from Biomethane

Similar to biomethane, other biofuels also exhibit better properties as compared to fossil fuels. Compression of biomethane yields bio-CNG, which exhibits similar properties in various applications such as industrial, transport, domestic, etc. Moreover, bio-CNG occupies a lesser volume.

Biomethane can also be converted to diesel, liquified petroleum (LPG) or methanol by subjecting it to a high pressure between 0.5 to 15 MP (Jeremiah et al. 2023). Biological conversion of biomethane to methanol can be carried out by methanotrophic bacteria and methane monooxygenase (Jeremiah et al. 2023).

6.7 Biomethane in Electricity Production

Biomethane is used for generation of electricity in Combined Heat and Power (CHP) plants, fuel cells, combustion engines, etc. with minimal environmental impact (Jeremiah et al. 2023). CHP plants generate electricity by burning fuel (natural gas/biogas) and then capturing the heat thus produced. This heat is then converted into thermal energy as steam or hot water. CHP systems may use biomethane as fuel in their four-stroke/diesel engines (Kaparaju and Rintala 2013). By upgrading biogas to biomethane, CHP systems utilize renewable energy while proposing waste heat for heating or industrial processes.

6.7.1 Agricultural Sector

Biomethane production from agricultural waste has environmental and economic benefits to farmers while contributing to renewable energy goals. Anaerobic digestion of livestock manure, crop residues, and organic by-products generates biogas, which can be upgraded to biomethane for use in energy generation or injection into natural gas grids. Biogas plants integrated with agriculture allows a proper means of waste management while also reducing methane emissions from manure storage. Co-digestion of various feedstocks is found to enhance biogas yields and process stability, optimizing resource utilization and revenue generation for farmers (Ghosh et al. 2020).

6.8 Conclusion

The production of biomethane not only offers a high-quality, versatile, and renewable source of energy that can be seamlessly integrated into the existing natural gas infrastructures but also provides a sustainable alternative to fossil fuels. Its applications extend to various areas, transportations being the most important one. The production methods of biomethane are highly flexible which allows for adaptations based on economic conditions and material availability, hence enhancing its viability across diverse geographic and industrial contexts.

Despite many advancements and the important future aspects of biomethane, this chapter also acknowledges the challenges that still remain, such as the need for improved pretreatment methods and more efficient bioconversion processes. Addressing these challenges will allow us to fully harness the potential of biomethane as the primary source of energy in the various applications described as well as more. Biomethane production through anaerobic digestion has multiple benefits as discussed. It improves waste management, contributes to sustainability, and reduces greenhouse gas emissions. Recent trends highlight a need to optimize the production process and enhance the efficiency of AD systems to expand the use of biomethane.

In conclusion, the transition to biomethane and other renewable energy sources represents a significant step towards a more sustainable and eco-friendly future. Further technological advancements and continued innovations are necessary to understand the full potential of biomethane and other biogases as primary sources of energy, ultimately helping in contributing to a sustainable global energy system.

References

A.V.S.L. Sai Bharadwaj, Subhabrata Dev, Jingshun Zhuang, Yunxuan Wang, Chang Geun Yoo, Byong-Hun Jeon et al. (2023). Review of chemical pretreatment of lignocellulosic biomass using low-liquid and low-chemical catalysts for effective bioconversion. Bioresource Technology 368: 128339, ISSN 0960-8524, https://doi.org/10.1016/j.biortech.2022.128339.

Abouelenien, F., Namba, Y., Kosseva, M. R., Nishio, N. and Nakashimada, Y. (2014). Enhancement of methane production from co-digestion of chicken manure with agricultural wastes. Bioresource Technology 159: 80–87. doi: 10.1016/j.biortech.2014.02.050.

Alves, H. J., Bley Junior, C., Niklevicz, R. R., Frigo, E. P., Frigo, M. S. and Coimbra-Araújo, C. H. (2013). Overview of hydrogen production technologies from biogas and the applications in fuel cells. pp. 5215–5225 *In*: International Journal of Hydrogen Energy. Vol. 38.

American Biogas Council, EPA LMOP (2020). Water Environment Federation "Enabling the Future," EPA AgSTAR 2023, EPA "Anaerobic Digestion Facilities Processing Food Waste in the United States" January 2021. https://americanbiogascouncil.org/biogas-market-snapshot/.

Amin, F. R., Khalid, H., Zhang, H., Rahman, S., Zhang, R., Liu, G. et al. (2017). Pretreatment methods of lignocellulosic biomass for anaerobic digestion. *In*: AMB Express (Vol. 7, Issue 1). Springer Verlag. https://doi.org/10.1186/s13568-017-0375-4.

Anacleto, T. M., Kozlowsky-Suzuki, B., Björn, A., Yekta, S. S., Masuda, L. S. M., de Oliveira, V. P. et al. (2024). Methane yield response to pretreatment is dependent on substrate chemical composition: a meta-analysis on anaerobic digestion systems. Scientific Reports 14(1). https://doi.org/10.1038/s41598-024-51603-9.

Arelli, V., Juntupally, S., Begum, S. and Anupoju, G. R. (2022). Solid state anaerobic digestion of organic waste for the generation of biogas and bio manure. Advanced Organic Waste Management: Sustainable Practices and Approaches 247–277. https://doi.org/10.1016/B978-0-323-85792-5.00023-X.

Saha, B. C., Iten, L. B., Cotta, M. A. and Wu, Y. V. (2005). Dilute acid pretreatment, enzymatic saccharification and fermentation of wheat straw to ethanol, Process Biochem. 40: 3693–3700.

Balat, M. (2011). Lignocellulosic biomass: a potential feedstock for bioethanol production. Energies 4(4): 919–966.

Baruah, J., Nath, B. K., Sharma, R., Kumar, S., Deka, R. C., Baruah, D. C. et al. (2018). Recent trends in the pretreatment of lignocellulosic biomass for value-added products. In Frontiers in Energy Research (Vol. 6, Issue DEC). Frontiers Media S.A. https://doi.org/10.3389/fenrg.2018.00141.

Braz, G. H. R., Fernandez-Gonzalez, N., Lema, J. M. and Carballa, M. (2018). The time response of anaerobic digestion microbiome during an organic loading rate shock. Applied Microbiology and Biotechnology 102(23): 10285–10297. https://doi.org/10.1007/s00253-018-9383-9.

Brodeur, G., Yau, E., Badal, K., Collier, J., Ramachandran, K. B. and Ramakrishnan, S. (2011). Chemical and physicochemical pretreatment of lignocellulosic biomass: A review. In Enzyme Research (Vol. 2011, Issue 1). https://doi.org/10.4061/2011/787532.

Bruni, E., Jensen, A. P. and Angelidaki, I. (2010). Steam treatment of digested biofibers for increasing biogas production. Bioresource Technology 101(19): 7668–7671. https://doi.org/10.1016/j.biortech.2010.04.064.

Calabrò, P. S., Elbeshbishy, E., Laqa Kakar, F. and Zema, D. A. (2024). A short bibliographic review concerning biomethane production from wastewater sludge. Waste Management & Research: The Journal of the International Solid Wastes and Public Cleansing Association, ISWA, 734242X241252906. Advance Online Publication. https://doi.org/10.1177/0734242X241252906.

Chai, J., Jiang, J., Gong, Y., Wu, P., Wang, A., Zhang, X. et al. (2023). Recent mechanistic understanding of Fischer-Tropsch synthesis on Fe-Carbide. Catalysts 13(7). doi:10.3390/catal13071052.

Chen, H., Liu, J., Chang, X., Chen, D., Xue, Y., Liu, P. et al. (2017). A review on the pretreatment of lignocellulose for high-value chemicals. Fuel Processing Technology 160: 196–206. https://doi.org/10.1016/J.FUPROC.2016.12.007.

Chundawat, S. P. S., Pal, R. K., Zhao, C., Campbell, T., Teymouri, F., Videto, J. et al. (2020). Ammonia Fiber Expansion (AFEX) pretreatment of lignocellulosic biomass. J. Vis. Exp. 2020 Apr 18; (158). doi: 10.3791/57488. PMID: 32364543.

Coarita Fernandez, H., Amaya Ramirez, D., Teixeira Franco, R., Buffière, P. and Bayard, R. (2020). Methods for the evaluation of industrial mechanical pretreatments before anaerobic digesters. Molecules 25(4): 860. https://doi.org/10.3390/molecules25040860.

Dahlgren, S. (2022). Biogas-based fuels as renewable energy in the transport sector: an overview of the potential of using CBG, LBG and other vehicle fuels produced from biogas. Biofuels 13(5): 587–599. doi:10.1080/17597269.2020.1821571.

Dong, M., Wu, Y., Li, Q., Tian, G., Yang, B., Li, Y. et al. (2015). Investigation of methanogenic community structures in rural biogas digesters from different climatic regions in Yunnan, southwest China. Current Microbiology 70(5): 679–684. https://doi.org/10.1007/s00284-015-0775-0.

Donkor, K. O., Gottumukkala, L. D., Lin, R. and Murphy, J. D. (2022). A perspective on the combination of alkali pre-treatment with bioaugmentation to improve biogas production from lignocellulose biomass. Bioresource Technology 351. https://doi.org/10.1016/j.biortech.2022.126950.

Frigon, J. C., Mehta, P. and Guiot, S. R. (2011). Impact of mechanical, chemical and enzymatic pre-treatments on the methane yield from the anaerobic digestion of switchgrass. Biomass Bioenerg. 36: 1–11. doi: 10.1016/j.biombioe.2011.02.013.

Galbe, M. and Wallberg, O. (2019). Pretreatment for biorefineries: a review of common methods for efficient utilisation of lignocellulosic materials. Biotechnol Biofuels 12: 294. https://doi.org/10.1186/s13068-019-1634-1.

Ghosh, P., Shah, G., Sahota, S., Singh, L. and Vijay, V. K. (2020). Biogas production from waste: Technical overview, progress, and challenges. pp. 89–104. *In*: Bioreactors: Sustainable Design and Industrial Applications in Mitigation of GHG Emissions. Elsevier.

Hagos, K., Zong, J., Li, D., Liu, C. and Lu, X. H. (2016). Anaerobic co-digestion process for biogas production: progress, challenges and perspectives. Renew. Sust. Energ. Rev. 76: 1485–1496. doi: 10.1016/j.rser.2016.11.184.

Hashemi, B., Sarker, S., Lamb, J. J. and Lien, K. M. (2021). Yield improvements in anaerobic digestion of lignocellulosic feedstocks. Journal of Cleaner Production 288. https://doi.org/10.1016/j.jclepro.2020.125447.

Herbes, C. (2017). Marketing of biomethane. pp. 151–171. *In*: Management for Professionals. Vol. Part F601. Springer Nature.

Herz, G., Reichelt, E. and Jahn, M. (2017). Design and evaluation of a Fischer-Tropsch process for the production of waxes from biogas. Energy 132: 370–381. doi:10.1016/j.energy.2017.05.102.

https://www.europeanbiogas.eu/wp-content/uploads/2019/07/Biomethane-in-transport.pdf.

Hupfauf, S., Plattner, P., Wagner, A. O., Kaufmann, R., Insam, H. and Podmirseg, S. M. (2018). Temperature shapes the microbiota in anaerobic digestion and drives efficiency to a maximum at 45°C. Bioresource Technology 269: 309–318. https://doi.org/10.1016/j.biortech.2018.08.106.

IEA. (2020), Outlook for biogas and biomethane: Prospects for organic growth, IEA, Paris https://www.iea.org/reports/outlook-for-biogas-and-biomethane-prospects-for-organic-growth, License: CC BY 4.0.

Janaun, J. and Ellis, N. (2010). Perspectives on biodiesel as a sustainable fuel. Renewable and Sustainable Energy Reviews 14(4): 1312–1320. https://doi.org/10.1016/j.rser.2009.12.011.

Jeremiah, M., Kabeyi, B. and Olanrewaju, O. A. (2022). Biomethane Production and Applications. www.intechopen.com.

Jeremiah, M., Kabeyi, B. and Olanrewaju, O. A. (2023). Diesel to gas engine power plant conversion: a review and preliminary design for an operating power plant. Research Article Journal of Energy Management and Technology (JEMT) 7: 3. doi:10.22109/JEMT.2022.292982.1312. http://dx.doi.org/10.22109/JEMT.2022.292982.1312.

Kaparaju, P. and Rintala, J. (2013). Generation of heat and power from biogas for stationary applications: Boilers, gas engines and turbines, combined heat and power (CHP) plants and fuel cells. pp. 404–427. *In*: The Biogas Handbook: Science, Production and Applications. Elsevier Inc.

Kheiralipour, K., Khoobbakht, M. and Karimi, M. (2024). Effect of biodiesel on environmental impacts of diesel mechanical power generation by life cycle assessment. Energy 289. https://doi.org/10.1016/j.energy.2023.129948.

Koonaphapdeelert, S., Aggarangsi, P. and Moran, J. (2020). Green Energy and Technology Biomethane Production and Applications. http://www.springer.com/series/8059.

Kouzi, A. I., Puranen, M. and Kontro, M. H. (2020). Evaluation of the factors limiting biogas production in full-scale processes and increasing the biogas production efficiency. Environmental Science and Pollution Research International 27(22): 28155–28168. https://doi.org/10.1007/s11356-020-09035-1.

Kumar, P., Barrett, D. M., Delwiche, M. J. and Stroeve, P. (2009). Methods for pretreatment of lignocellulosic biomass for efficient hydrolysis and biofuel production. pp. 3713–3729. *In*: Industrial and Engineering Chemistry Research (Vol. 48, Issue 8). https://doi.org/10.1021/ie801542g.

Laiq Ur Rehman, M., Iqbal, A., Chang, C. C., Li, W. and Ju, M. (2019). Anaerobic digestion. Water Environment Research: A Research Publication of the Water Environment Federation 91(10): 1253–1271. https://doi.org/10.1002/wer.1219.

Liebetrau, J., Sträuber, H., Kretzschmar, J., Denysenko, V. and Nelles, M. (2017). Anaerobic digestion. *In*: Wagemann, K. and Tippkötter, N. (eds.). Biorefineries. Advances in Biochemical Engineering/Biotechnology, vol 166. Springer, Cham. https://doi.org/10.1007/10_2016_67.

Liu, C. M., Wachemo, A. C., Tong, H., Shi, S. H., Zhang, L., Yuan, H. R. et al. (2018). Biogas production and microbial community properties during anaerobic digestion of corn stover at different temperatures. Bioresource Technology 261: 93–103. https://doi.org/10.1016/j.biortech.2017.12.076.

Liu, Z., Si, B., Li, J., He, J., Zhang, C., Lu, Y. et al. (2018). Bioprocess engineering for biohythane production from low-grade waste biomass: Technical challenges towards scale up. Current Opinion in Biotechnology 50: 25–31. https://doi.org/10.1016/j.copbio.2017.08.014.

Lohani, S. P. and Havukainen, J. (2018). Anaerobic digestion: factors affecting anaerobic digestion process. *In*: Varjani, S., Gnansounou, E., Gurunathan, B., Pant, D. and Zakaria, Z. (eds.). Waste Bioremediation. Energy, Environment, and Sustainability. Springer, Singapore. https://doi.org/10.1007/978-981-10-7413-4_18.

Lozanovski, A., Linder, J. P. and Bos, U. (2014). Environmental evaluation and comparison of selected industrial scale biomethane production facilities across Europe. Int. J. Life Cycle Assess 19: 1823–1832.

Mackie, K. L., Brownell, H. H., West, K. L. and Saddler, J. N. (1985). Effect of sulphur dioxide and sulphuric acid on steam explosion of aspenwood. Journal of Wood Chemistry and Technology 5(3): 405–425. https://doi.org/10.1080/02773818508085202.

Mao, C., Zhang, T., Wang, X., Feng, Y., Ren, G. and Yang, G. (2017). Process performance and methane production optimizing of anaerobic co-digestion of swine manure and corn straw. Sci. Rep. 2017 Aug 24; 7(1): 9379. doi: 10.1038/s41598-017-09977-6. PMID: 28839230; PMCID: PMC5571176.

Mendes, Carlos, Ferreira, Naiara, Furtado, Cristina and Maria, Furtado. (2015). Isolation and characterization of nanocrystalline cellulose from corn Husk. Materials Letters 148. 10.1016/j.matlet.2015.02.047.

Meyer, T. and Edwards, E. A. (2014). Anaerobic digestion of pulp and paper mill wastewater and sludge. Water Research 65: 321–349. https://doi.org/10.1016/j.watres.2014.07.022.

Mourshed, M., Kamal, M., Masuk, N. I., Chowdhury, S. A. and Masud, M. H. (2023). Anaerobic digestion process of biomass. Reference Module in Earth Systems and Environmental Sciences. https://doi.org/10.1016/B978-0-323-93940-9.00025-6.

Musa, M. A., Idrus, S., Hasfalina, C. M. and Daud, N. N. N. (2018). Effect of organic loading rate on anaerobic digestion performance of mesophilic (UASB) reactor using cattle slaughterhouse wastewater as substrate. International Journal of Environmental Research and Public Health 15(10): 1–19.

Nguyen, L. N., Kumar, J., Vu, M. T., Mohammed, J. A. H., Pathak, N., Commault, A. S. et al. (2021). Biomethane production from anaerobic co-digestion at wastewater treatment plants: A critical review on development and innovations in biogas upgrading techniques. The Science of the Total Environment 765: 142753. https://doi.org/10.1016/j.scitotenv.2020.142753.

Nouhou Moussa, A. W., Sawadogo, B., Konate, Y., Thianhoun, B., Sidibe, S. D. S. and Heran, M. (2023). Influence of solid retention time on membrane fouling and biogas recovery in anerobic membrane bioreactor treating sugarcane industry wastewater in sahelian climate. Membranes 13(8): 710. https://doi.org/10.3390/membranes13080710.

Pääkkönen, A., Aro, K., Aalto, P., Konttinen, J. and Kojo, M. (2019). The potential of biomethane in replacing fossil fuels in heavy transport—a case study on Finland. Sustainability (Switzerland) 11(17). doi:10.3390/su11174750.

Pang, F., Xue, S., Yu, S., Zhang, C., Li, B. and Kang, Y. (2013). Effects of combination of steam explosion and microwave irradiation (SE-MI) pretreatment on enzymatic hydrolysis, sugar yields and structural properties of corn stover. Industrial Crops and Products 42(1): 402–408. https://doi.org/10.1016/j.indcrop.2012.06.016.

Park, C., Lee, C., Kim, S., Chen, Y. and Chase, H. A. (2005). Upgrading of anaerobic digestion by incorporating two different hydrolysis processes. Journal of Bioscience and Bioengineering 100(2): 164–167. https://doi.org/10.1263/jbb.100.164.

Pavičić, J., Novak Mavar, K., Brkić, V. and Simon, K. (2022). Biogas and biomethane production and usage: technology development, advantages and challenges in Europe. Energies 15: 2940. https://doi.org/10.3390/en15082940.

Rahimi, N., Azarpour, A., Talebi, M. and Rezaei, M. (2018). Pretreatment of lignocellulosic biomass for enhanced biogas production: A review. Journal of Environmental Management 228.

Röder, M. and Welfle, A. (2018). Bioenergy. Managing Global Warming: An Interface of Technology and Human Issues 379–398. https://doi.org/10.1016/B978-0-12-814104-5.00012-0.

Sun, Y. and Cheng, J. (2002). Bioconversion of lignocellulosic biomass: biochemical and molecular perspectives. Biotechnology Advances 20(2): 93–120.

Thanh, P. M., Ketheesan, B., Yan, Z. and Stuckey, D. (2016). Trace metal speciation and bioavailability in anaerobic digestion: A review. Biotechnology Advances 34(2): 122–136. https://doi.org/10.1016/j.biotechadv.2015.12.006.

Toscan, A., Morais, A. R. C., Paixão, S. M., Alves, L., Andreaus, J., Camassola, M. et al. (2017). High-pressure carbon dioxide/water pre-treatment of sugarcane bagasse and elephant grass: Assessment of the effect of biomass composition on process efficiency. Bioresource Technology 224: 639–647. https://doi.org/10.1016/J.BIORTECH.2016.11.101.

Tsigkou, K., Zagklis, D., Parasoglou, M., Zafiri, C. and Kornaros, M. (2022). Proposed protocol for rate-limiting step determination during anaerobic digestion of complex substrates. Bioresource Technology 361: 127660. https://doi.org/10.1016/j.biortech.2022.127660.

Van Lier, Jules B., Nidal Mahmoud and Grietje Zeeman. (2008). Anaerobic wastewater treatment. Biological Wastewater Treatment: Principles, Modeling and Design 415–456.

Vu, H. P., Nguyen, L. N., Wang, Q., Ngo, H. H., Liu, Q., Zhang, X. et al. (2022). Hydrogen sulphide management in anaerobic digestion: A critical review on input control, process regulation, and post-treatment. Bioresource Technology 346: 126634. https://doi.org/10.1016/J.BIORTECH.2021.126634.

Wang, Z., Wu, G. and Jönsson, L. J. (2018). Effects of impregnation of softwood with sulfuric acid and sulfur dioxide on chemical and physical characteristics, enzymatic digestibility, and fermentability. Bioresource Technology 247: 200–208. https://doi.org/10.1016/J.BIORTECH.2017.09.081.

Yang, J., Tang, S., Mei, W., Chen, Y., Yi, W., Lv, P. et al. (2024). Valorising lignocellulosic biomass to high-performance electrocatalysts via anaerobic digestion pretreatment. Biochar 6(1). https://doi.org/10.1007/s42773-024-00311-8.

Zha, Y., Westerhuis, J. A., Muilwijk, B., Overkamp, K. M., Nijmeijer, B. M., Coulier, L. et al. (2014). Identifying inhibitory compounds in lignocellulosic biomass hydrolysates using an exometabolomics approach. BMC Biotechnology 14: 22. https://doi.org/10.1186/1472-6750-14-22.

Zhang, R. and El-Mashad, H. M. (2012). Biogas production from co-digestion of dairy manure and food waste. Bioresource Technology 123: 535–542.

Zou, S., Wang, X., Chen, Y., Wan, H. and Feng, Y. (2016). Enhancement of biogas production in anaerobic co-digestion by ultrasonic pretreatment. Energy Conversion and Management 112: 226–235. doi: 10.1016/j.enconman.2015.12.087.

Chapter 7

Micro-Aeration in Anaerobic Microbial Communities

Carla Pereira Magalhães,[1] *Luís Costa,*[1] *Maria Salomé Duarte,*[1,2]
Maria Alcina Pereira[1,2] and *Ana Júlia Cavaleiro*[1,2,*]

7.1 Introduction

Energy is essential for life. In today's world, generating energy from fossil fuels has a significant environmental impact. Therefore, there is an urgent need to shift towards more sustainable and clean energy sources (European Biogas Association 2024). Anaerobic Digestion (AD) stands up as sustainable technology for producing biogas, which can serve as a renewable energy source and help reduce greenhouse gas emissions. Biogas can be upgraded to biomethane and used as transport fuel or directly injected in the natural gas grid, thus contributing for the mix of renewables in Europe (European Biogas Association 2024). A great variety of complex organic wastes can be used in AD as substrate, such as domestic and industrial wastewater, food waste, animal manure and municipal sewage sludge, among others (Fig. 7.1).

In recent years, micro-aeration has been described as a promising technique to improve AD processes. Unlike traditional anaerobic processes, that operate strictly in the absence of oxygen, micro-aeration involves the introduction of small amounts of oxygen, either as a pre-treatment or directly into the anaerobic digester. This strategy aims to optimize microbial activity and enhance process performance while maintaining the overall anaerobic conditions necessary for biogas production (Johansen and Bakke 2006, Lim and Wang 2013).

[1] CEB - Centre of Biological Engineering, University of Minho, Campus de Gualtar, Braga, Portugal.
[2] LABBELS – Associate Laboratory, Braga/Guimarães, Portugal.
Emails: c.pereira.mg@gmail.com, id11211@alunos.uminho.pt, salomeduarte@ceb.uminho.pt, alcina@deb.uminho.pt
* Corresponding author: acavaleiro@ceb.uminho.pt

DOI: 10.1201/9781003506898-7

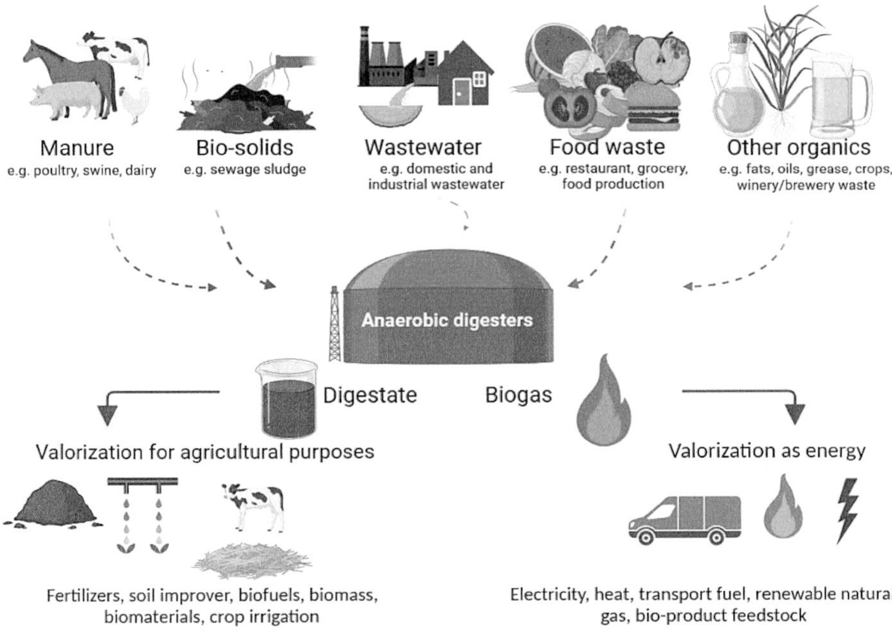

Figure 7.1. Possible feedstocks for anaerobic digestion and potential applications of its mains products (biogas and digestate). Created in BioRender. Magalhães, C. (2025). https://BioRender.com/s23n082.

The application of oxygen in AD processes was considered an antagonism, because anaerobes cannot tolerate high oxygen doses as it poisons their core strategies of energy production (Thauer 2019, Khademian and Imlay 2020). Methanogens and other so called obligate anaerobes were thought to be deprived of strategies to deal with oxidative stress (Khademian and Imlay 2021), but this idea has been recently surpassed, and several enzymes are currently known to scavenge oxidative radicals and control oxidative stress in these microorganisms (Lu and Imlay 2021). Instead of causing immediate inhibition, micro-aeration can enhance the performance of AD processes, by aiding to surpass some process limitations – e.g., hydrolysis of lignocellulosic and other solid wastes, Volatile Fatty Acids (VFA) accumulation, H_2S accumulation which is toxic and corrosive, and low methane yields (Johansen and Bakke 2006, Nguyen and Khanal 2018, Magdalena et al. 2022). Micro-aeration promotes the simultaneous occurrence of both anaerobes and micro-aerobes, thereby augmenting the activity and diversity of the microbial communities involved in AD processes (Nguyen and Khanal 2018).

7.2 Definition of Micro-Aerobic Environment

Various terms are employed across scientific areas to define oxygen-deficient conditions. Recently, Berg and collaborators have summarized existent terminology, based on geochemical criteria and oxygen upper thresholds (Berg et al. 2022). Among the different biological regimes identified, micro-aerobic conditions were ascribed to oxygen concentrations lower than 10 μmol L^{-1}, with the extreme anaerobic and aerobic regimes being associated with oxygen upper thresholds of 0 and 61–260 μmol L^{-1}, respectively. Nanaerobic is ascribed to the ability of certain prokaryotes to grow with oxygen concentrations as low as nanomolar levels (< 1 μmol L^{-1}) (Berg et al. 2022).

Micro-aeration-based AD create a unique habitat that combines both anaerobic and aerobic conditions, thus preserving beneficial niches for both anaerobes and micro-aerobes (Nguyen and Khanal 2018). In anaerobic conditions, organic materials are only partially converted into energy-rich molecules. Due to a lack of an effective mechanism for electron transport, anaerobes produce less energy from organic degradation reactions. In opposition, aerobic conditions promote rapid consumption of organic matter and microbial growth, facilitated by a highly active metabolism (Nguyen and Khanal 2018).

7.3 Micro-Aeration in AD

Micro-aeration in AD is characterized by the introduction of small concentrations of oxygen into an anaerobic digester, typically ranging from 0.005 to 5 L L^{-1} d^{-1} (expressed as volume of O_2 per unit volume of reactor per day).

Micro-aeration is a versatile and effective strategy for enhancing various components of AD, offering benefits across different stages of the process. Primarily, it accelerates hydrolysis, the pivotal initial step in AD. Hydrolysis is generally conducted by extracellular hydrolytic enzymes, which hydrolyze complex substrates to simple organic molecules, and it is, in most cases, the process's limiting step. Increasing the activity of the microorganisms involved in this stage is the key to facilitate these reactions. Micro-doses of oxygen promote the diversity and abundance of fast-growing Facultative Anaerobic Bacteria (FAB), and consequently, the production of hydrolytic enzymes (e.g., cellulases, xylanases, lipases). Hydrolysis of various substrates has been successfully promoted by micro-aeration, either as pre-treatment or with direct O_2 administration in the reactors. This approach has been tested with sewage sludge (Montalvo et al. 2016, 2018, Hasegawa et al. 2000), food waste (Xu et al. 2014, Zhen et al. 2020), lignocellulosic material (Nguyen et al. 2019, Xu et al. 2021), and others. Since FAB may serve multiple functions, acidogenesis is also frequently stimulated, with the generation of VFA (Jagadabhi et al. 2010, Lim and Wang 2013, Lim et al. 2014). An active community will keep the substrates flowing, since acetogenic bacteria use different organic acids to produce acetate. Tsapekos and colleagues showed a rapid accumulation of VFA (mostly acetate) after an oxygen pulse in the degradation of wheat straw (Tsapekos et al. 2017). Facilitation of hydrolysis, acidogenesis and acetogenesis under micro-aeration lead to AD processes with higher stability and lower lag phases (Wu et al. 2021, Cao et al. 2022).

Additionally, micro-aeration mitigates hydrogen sulfide (H_2S) accumulation, a common challenge in biogas production, thereby safeguarding infrastructure integrity and optimizing methane yields. During the AD process, sulfur compounds present in the substrate are converted by Sulfate-Reducing Bacteria (SRB) into hydrogen sulfide (H_2S), which is corrosive and harmful resulting in serious economic losses. Additionally, sulfate-reducing bacteria (SRB) also compete with methanogens for carbon and electron sources (e.g., H_2, CO_2, and acetate), lowering methane production yields (Krayzelova et al. 2015, Dai et al. 2017). Using micro-aeration for biogas desulfurization can be done directly in the reactor or through recirculation of sludge or biogas (Chen et al. 2017, Jeníček et al. 2017). Compared with other techniques, micro-aeration has high removal efficiencies (up to 99%) and low operational costs (Fdz.-Polanco et al. 2009, Ramos et al. 2014).

Recently, micro-aeration has been applied for the removal of persistent micro-/pollutants, e.g., chlorine phenol, aniline, petrochemical and pharmaceutical wastewater, enhancing the quality of the treated water and reducing environmental impact (Zhang et al. 2021). As an example, Wu and co-workers evaluated the effect of micro-aeration on petrochemical wastewater, by applying dissolved oxygen concentrations of 0.2–0.3 mg L^{-1} (average Oxidation-Reduction Potential (ORP) of –210 mV), and verified that the wastewater's toxicity was greatly reduced (Wu et al. 2015). Further, in a full-scale system, the same authors showed that the insertion of limited aeration in a hydrolysis-acidification tank (as pre-treatment before the anoxic tank) was efficient in the degradation of 49 micro-pollutants of a petrochemical wastewater, e.g., BTEX (Benzene, Toluene, Ethylbenzene and Xylene) and phenols (Wu et al. 2018). Duarte and collaborators applied micro-aeration in the AD of lipid-rich wastewater, which facilitated the conversion of long chain fatty acids (the products of lipids' hydrolysis) to methane, potentiating the anaerobic treatment of this type of waste (Duarte et al. 2018).

Several factors can influence/determine the micro-aeration rate and efficiency in AD processes, namely, the reactor configuration, the micro-aeration method, the inoculum, the substrate, the organic loading rate, and the micro-aeration purpose (Nguyen and Khanal 2018, Chen et al. 2020). According to that, numerous techniques can be used to create a micro-aerobic environment. The dosing of air or pure oxygen in the liquid phase, in the gaseous phase and in the biogas recirculation are the most common injection techniques used in anaerobic digesters (Ramos et al. 2014, Krayzelova et al. 2015). Another innovative micro/nano-aeration technology is the use of nanobubbles (50–200 nm diameter), allowing higher oxygen transfer rate and efficiency. Nanobubbles are stable in water for days or even months (Chuenchart et al. 2021). Depending on the practical requirements, aeration frequencies might be a single dose (pulse), intermittent, or continuous (Lim et al. 2014, Tsapekos et al. 2017, Nguyen and Khanal 2018), and be added in different stages of the AD process, as pre-treatment, directly in the anaerobic reactor or as a post-process (Girotto et al. 2018).

The control method employed in AD technologies with micro-aeration is critical. Oxygen levels may be too low to influence the process and, as a result, the methane output will not benefit from micro-aeration (Xu et al. 2014). On the other hand, the application of excessive oxygen levels may result in oxidation of easily available substrates, or methane consumption by aerobic methanotrophs. In addition, anaerobic methanogenic archaea can be inhibited, resulting in lower levels of methane or system failure (Zhu et al. 2009, Botheju and Bakke 2011).

7.3.1 Oxidation-Reduction Potential (ORP) as Micro-Aeration Control Parameter

The trace amounts of oxygen in micro-aerated systems are usually not detectable by common dissolved oxygen probes, but Oxidation-Reduction Potential (ORP) can be monitored instead and used to detect even minute changes in oxygen concentration. ORP values of –50 mV (with reference to standard hydrogen reference electrode) are associated to dissolved oxygen concentrations of 0.1 mg L^{-1} (3 µmol L^{-1}) (Liu et al. 2013, Nguyen and Khanal 2018). Micro-aerobic conditions in aqueous media correspond to ORP values ranging from 0 to –300 mV, whereas average ORP levels below –300 mV are associated with anaerobic conditions; ORP values greater than 0 mV indicate aerobic conditions (Khanal and Huang 2006, Krayzelova et al. 2015)—Fig. 7.2. Anaerobic microbial activity may proceed throughout a wide range of ORP values, although exhibiting preferential ORP levels—e.g., anaerobic proton reduction ($2H^+/H_2$) is dominant below –300/–400 mV, and thus methanogenesis (CO_2/CH_4) occur preferentially over values close to –300 mV, while ORP values above –200 mV benefit nitrification (NH_4^+/NO_3^-) and fermentation processes (Nguyen and Khanal 2018).

The ORP is a chemical measurement of electrochemical potential at zero current that shows the status of competing electron transfer reactions in solution. It is determined by a ratio of oxidative to reductive components and indicates the overall balance of these components at a given period, where an oxidation reaction includes a loss of one or more electrons, while reduction reaction entails the gain of electrons (Liu et al. 2017). As external redox potential correlates to the net balance of intracellular reducing equivalents and the changes in the cellular redox environment, ORP also reflects the metabolic activity of microorganisms (Liu et al. 2017).

Several studies have used ORP as an aeration control parameter in AD processes treating diverse compounds, such as synthetic wastewater (Zitomer and Shrout 1998), brown water (Lim et al. 2014), swine wastewater (Yang et al. 2021), petrochemical wastewater (Wu et al. 2015), wheat straw (Tsapekos et al. 2017) or napier grass (Nguyen et al. 2019). The establishment of an ORP window, in which individual metabolic processes could occur without the disruption of methanogenesis, is the goal. However, the ORP in AD systems is also affected by other operating factors (e.g., pH, temperature, ionic strength), and thus, only if measured under

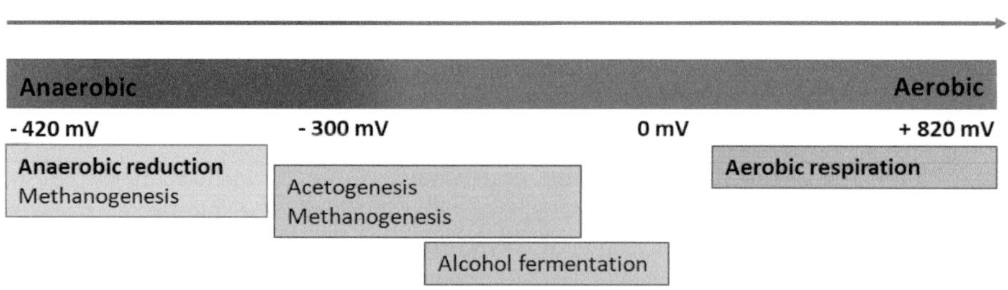

Figure 7.2. Oxidation-reduction potentials (ORP) of biochemical processes. Processes highlighted in bold can be dominant at specific ORP values.

controlled settings, ORP can be an effective parameter for monitoring and managing micro-aeration in AD systems.

7.4 Influence of Micro-Aeration in Anaerobic Microbial Communities

In AD systems, microbial communities are composed by diverse microorganisms which possess different affinity/sensibility to oxygen, each one performing different functions that may be essential for an efficient process (Stackebrandt 2004). These include:

i) strictly aerobic microorganisms, which utilize oxygen as the terminal electron acceptor for energy production;

ii) facultative anaerobes, that are capable of respiring with oxygen but can switch to fermentation or anaerobic respiration in its absence. Many fermenters fall within this category. These microorganisms are thought to have a variety of roles in AD, including improving the process's initial steps (e.g., hydrolysis) and contributing to more effective fatty acid breakdown (Botheju and Bakke 2011);

iii) oxygen-tolerant microorganisms, that are able to survive in the presence of oxygen but undergo fermentation;

iv) micro-aerophilic microorganisms, that require oxygen at O_2 concentrations lower than atmospheric levels (typically 2–10% O_2) for optimal growth;

v) strictly anaerobic microorganisms, that perform anaerobic respiration or fermentation, and for which oxygen may be toxic (Berg et al. 2022). This group includes methanogens that are strictly anaerobic archaea.

7.4.1 Intrinsic Mechanisms of Oxidative Resistance in Methanogens and Strict Anaerobes

Oxygen and Reactive Oxygen Species (ROS), such as superoxide (the primary oxygen free radical - $O^{\bullet}{}_2{}^{-}$), hydrogen peroxide (H_2O_2), or hydroxyl radicals ($HO^{\bullet}$), are highly oxidative species capable of damaging proteins, DNA, and lipids, fundamental molecules in biochemistry (Khademian and Imlay 2021, Li et al. 2022). Initially, it was believed that methanogens and other so-called obligate anaerobic bacteria originated from branches of the tree of life that had never been exposed to oxygen, thus developing without strategies to cope with oxidative stress (Khademian and Imlay 2021). However, further investigations have revealed that strict anaerobic organisms also possess mechanisms to counteract these conditions. Physiological, ecological, and genomic analyses have uncovered antioxidant features in several anaerobes, indicating their capability to withstand oxidative stress (Horne and Lessner 2013, Lyu and Lu 2018, Khademian and Imlay 2021).Various enzymes have been identified in these microorganisms, that scavenge oxidative radicals and regulate oxidative stress (Jasso-Chávez et al. 2015, Lyu and Lu 2018, Lu and Imlay 2021), allowing them to tolerate brief exposure to oxygen. The advent of Next-Generation Sequencing (NGS) techniques has expedited the identification of the genes encoding

Table 7.1. ROS generation and O_2/ROS elimination reactions.

ROS generation reactions	
$O_2 + e^- \rightarrow O^{\bullet}{}_2^-$	
$2O_2^- + 2H^+ \rightarrow H_2O_2 + O_2$	
$O_2^- + H_2O_2 \rightarrow OH^{\bullet} + OH^- + O_2$	
$H_2O_2 + Fe^{2+} \rightarrow Fe^{3+} + OH^{\bullet} + OH^-$	
O_2/ROS elimination reactions	**Oxidative enzymes**
$O^{\bullet}{}_2^- + 2H^+ \rightarrow H_2O_2 + O_2$	SOD
$2H_2O_2 \rightarrow H_2O + O_2$	KAT
$O^{\bullet}{}_2^- + 2H^+ \rightarrow H_2O_2$	SOR [(a)]
O_2 + 2Ubiquinol-8 + $4H^+ \rightarrow$ Ubiquinone-8 + $2H_2O$	Cyt bd [(a)]

[(a)] Energy dependent

for those enzymes within the genomes of microorganisms, such as catalase (KAT), superoxide dismutase (SOD), superoxide reductase (SOR), peroxidase (PX), among others (Table 7.1). For instance, enzymes responsible for detoxifying oxygen have been identified in methanogens (such as *Methanobacterium formicicum* and *Methanospirillum hungatei*) and syntrophic bacteria (including *Syntrophomonas wolfei* and *S. zehnderi*) (Magalhães et al. 2023).

Aerotolerance is highly variable among anaerobes, and their behavior is not only justified by the variability of oxidative defense systems (O_2/ROS elimination) but also by the mechanisms of ROS production, and by self-repairing systems. Lyu and collaborators performed a functional analysis of methanogens genes involved in these strategies and determined two different clusters of methanogens – Class I (Methanobacteriales, Methanococcales and Methanopyrales) and Class II (Methanocellales, Methanomicrobiales and Methanosarcinales). Class II methanogens have evolved features that enable a better adaptation to oxidative environments, such as the usage of enzymes producing less ROS or enhancing ROS removal. This class presents a methanogenesis pathway based on a cytochrome/methanophenazyne mediated electron transfer mechanism which reduced $HO^{\bullet}$. It has also an enriched and diversified gene network encoding DNA/lipids/protein/Fe clusters base repair reduction enzymes (Lyu and Lu 2018). Notably, these diverse features can be found in a various range of microorganisms, demonstrating that nature evolved alternative tactics to suppress ROS stress.

Despite the intrinsic resistance that each microorganism may have to oxygen and ROS, this can potentially be adapted and changed over time. Works have shown that, as incubations were performed over time, shorter adaptation times and faster methane production rates could be attained, reaching adaptable cultures with higher resistance to oxygen. For example, air-adapted *Methanosarcina acetivorans* after 6 months under micro-aeration had similar methane production rates and 50 times lower production of detoxification proteins than non-adapted strain (Jasso-Chávez et al. 2015). *Desulfovibrio vulgaris* Hildenborough after 114 generations of growth under oxidative stress were able to grow at oxygen concentrations 32 times higher than the ancestral strain and conserve energy from this process (Schoeffler et al. 2019).

Regardless of having oxidation protective strategies and being adaptable, these microorganisms still cannot tolerate oxygen levels far from their native habitats and might be somehow damaged when in contact with oxygen. Information on oxygen toxicity and inhibition is still controversial, but as obligate anaerobes become genetically accessible, and O_2-sensitive enzymes physiologies and biochemical mechanisms are investigated, knowledge is acquired faster to understand the effects of O_2 on these microorganisms.

7.4.2 Other Strategies of Oxidative Resistance in Anaerobic Communities

The so-called strict anaerobes can not only survive in environments with oxygen but also may have their activity stimulated, a phenomenon known as the methane paradox. Studies carried out by Angle et al. (2017) challenged the paradigm that methanogenesis can only occur in anoxic habitats by demonstrating nine times more methanogenic activity in oxygenated soils compared to anoxic soils. These findings challenge a widely held assumption about methanogenesis and suggest that such oxygen tolerance is likely provided by other members of the community or by the maintenance of anoxic niches (Angle et al. 2017).

The mechanisms by which methanogens withstand oxygen stress are crucial for practical applications of micro-aeration technology. In addition to the intrinsic tolerance of certain methanogenic species to oxygen exposure, potential strategies include the low solubility of oxygen in water, rapid consumption of dissolved oxygen by facultative anaerobes, and the formation of microbial aggregates. Additionally, chemical compounds such as sulfide, cysteine or glutathione are also able to scavenge oxidative radicals. Depending on the oxygen dispersion method, the solubility of oxygen in water is low, reaching only about 398 $\mu mol\ L^{-1}$ at maximum saturation at 21°C (Fondriest Environmental 2013). Despite high oxygen loads during micro-aeration, dissolved oxygen levels in the liquid phase remain low, with minimal oxygen content capable of affecting obligate methanogens.

Microbial aggregates, like granules or biofilms, are typical in AD systems, consisting of a methanogen-dominated core and a bacteria-dominated surface. FAB, which include hydrolytic and fermentative bacteria, reside in the outer layers of these aggregates. Their consumption of oxygen serves to protect the strict anaerobes situated in the inner core from oxygen exposure (Botheju and Bakke 2011). This spatial organization results in decreasing oxygen levels from the outer layer to the inner core, until reaching the strict anaerobic tolerance limit. Additionally, a substrate gradient is established, where hydrolytic bacteria degrade complex organic matter into various intermediates through oxygen consumption. These intermediates then serve as substrates for fermentative bacteria and methanogens.

Even in small amounts, oxygen can potentially be used as an electron acceptor. This imposes a reorganization of metabolic pathways which causes a specific community shift. Particularly, FAB growth is stimulated under micro-aerobic conditions. Facultative anaerobes play a fundamental role in complex microbial communities within AD environments. They swiftly consume dissolved oxygen, facilitating the establishment of low redox potentials essential for obligate methanogens (Zitomer and Shrout 1998, Fu et al. 2015). These organisms also play an important role in establishing anaerobic conditions by neutralizing oxidative species (Fu et al. 2015).

In fact, when studying complex microbial communities, Morais et al. (2024) showed that methanogenic hydrogenotrophic and aceticlastic archaea were inhibited by oxygen concentrations higher than 0.5%, reaching complete inhibition at 5% O_2. However, in the presence of a substrate as ethanol, significant inhibition of the methane production by the anaerobic community (> 36%) was only observed for oxygen concentrations higher than 2.5%. Ethanol favored the activity of aerobic bacteria (AB)/FAB, that enhanced oxygen removal and helped to protect the activity of the microbial community (and in particular of the syntrophic consortia involved in ethanol conversion to methane) under micro-aerobic conditions. This work highlights the importance of AB/FAB in maintaining functional and resilient syntrophic communities, which is relevant for real AD systems, in which vestigial O_2 amounts are frequently present, as well as for AD systems using micro-aeration as a process strategy.

Similar results were obtained by Magalhães et al. (2023), in assays with constructed syntrophic consortia and using medium- and long-chain fatty acids as a substrate. These authors showed that the syntrophic activity was also inhibited at oxygen concentrations higher than 0.5%, but when in the presence of *Pseudomonas* spp. (working as model FAB), the syntrophic cultures successfully converted the fatty acids to methane with up to 2% O_2. These works have shown that the presence of FAB have a shielding effect against oxygen toxicity towards methanogens and other strict anaerobes, and contribute to more resilient anaerobic communities, able to thrive in the presence of oxygen and successfully convert organic compounds to methane.

An alternative role for FAB was suggested by Zabranska and Pokorna (2018). These authors propose that FAB may replace syntrophic bacteria, producing H_2 and CO_2 from fatty acids, thus working in direct sinergy with hydrogenotrophic methanogens, that consume H_2/CO_2. The combination of these biochemical processes is energetically more favorable than classical syntrophic reactions. The shift of dominant methanogenic pathway, from aceticlastic to hydrogenotrophic, following micro-aeration has been reported in several studies (as reviewed by Nguyen and Khanal 2018), and was previously assigned to the fact that hydrogenotrophic methanogens are generally more resilient to inhibitors than the aceticlastic methanogens. The possible role of FAB on these processes, as suggested by Zabranska and Pokorna (2018), open new perspectives of microbial interactions in AD and should be further studied.

7.5 Conclusions and Perspectives

Micro-aeration has increasingly gained significance in AD processes due to its various benefits, ultimately leading to enhanced methane production and process stability. However, its application often relies on a trial and error approach, reflecting a limited understanding of the underlying mechanisms and pathways.

A comprehensive understanding of how micro-aeration impacts microbial communities' composition, diversity and stability, as well as its influence on key metabolic pathways, is essential for revealing how it can foster synergistic interactions among facultative bacteria, anaerobic bacteria, and methanogens, thereby enhancing the overall performance of micro-aerated AD systems. Additionally, the long-term effects of micro-aeration on the adaptation of strict anaerobes and on shaping/driving complex microbial communities warrant further investigation.

To advance in this field, the co-occurrence of aerobic and anaerobic microbial respiration at both individual and community levels, as well as the biochemical mechanisms regulating the switching from aerobic to anaerobic respiration, also require further studies. This knowledge is important for sustaining simultaneous aerobic and anaerobic processes, harnessing the benefits of micro-aeration without jeopardizing the anaerobic process and avoid diverting an important fraction of the carbon flux from biomethane.

7.6 Acknowledgements

This study was supported by the Portuguese Foundation for Science and Technology (FCT) under the scope of the strategic funding of UID/04469: Centro de Engenharia Biológica da Universidade do Minho (CEB-UM/UM). The work of C.P. Magalhães was funded by FCT through the fellowships SFRH/BD/132845/2017 and COVID/BD/152450/2022. M.S. Duarte acknowledge FCT for the Junior Research contract under the scope of the Scientific Stimulus Employment 2022 (ref: 2022.06569. CEECIND/CP1718/CT0004; DOI 10.54499/2022.06569.CEECIND/CP1718/CT0004). L. Costa received research support from FCT through the fellowships 2023.01617.BD. The authors also acknowledge the Science Communication Office from LABBELS (LA/P/0029/2020).

References

Angle, J. C., Morin, T. H., Solden, L. M., Narrowe, A. B., Smith, G. J., Borton, M. A. et al. (2017). Methanogenesis in oxygenated soils is a substantial fraction of wetland methane emissions. Nat. Commun. 8(1): 1567. https://doi.org/10.1038/s41467-017-01753-4.

Berg, J. S., Ahmerkamp, S., Pjevac, P., Hausmann, B., Milucka, J. and Kuypers, M. M. M. (2022). How low can they go? Aerobic respiration by microorganisms under apparent anoxia. FEMS Microbiol. Rev. 46(3): fuac006. https://doi.org/10.1093/femsre/fuac006.

Botheju, D. and Bakke, R. (2011). Oxygen effects in anaerobic digestion—a review. Open Waste Manag. J. 4: 119. https://doi.org/10.2174/1876400201104010001.

Cao, Q., Zhang, W., Lian, T., Wang, S., Yin, F., Zhou, T. et al. (2022). Roles of micro-aeration on enhancing volatile fatty acids and lactic acid production from agricultural wastes. Bioresource Technol. 347: 126656. https://doi.org/10.1016/j.biortech.2021.126656.

Chen, C., Zhang, R.-C., Xu, X.-J., Fang, N., Wang, A.-J., Ren, N.-Q. et al. (2017). Enhanced performance of denitrifying sulfide removal process at high carbon to nitrogen ratios under micro-aerobic condition. Bioresource Technol. 232: 417–422. https://doi.org/10.1016/j.biortech.2017.02.031.

Chen, Q., Wu, W., Qi, D., Ding, Y. and Zhao, Z. (2020). Review on microaeration-based anaerobic digestion: state of the art, challenges, and prospectives. Sci. Total Environ. 710: 136388. https://doi.org/10.1016/j.scitotenv.2019.136388.

Chuenchart, W., Karki, R., Shitanaka, T., Marcelino, K. R., Lu, H. and Khanal, S. K. (2021). Nanobubble technology in anaerobic digestion: a review. Bioresour Technol. 329: 124916. https://doi.org/10.1016/j.biortech.2021.124916.

Dai, X., Hu, C., Zhang, D. and Chen, Y. (2017). A new method for the simultaneous enhancement of methane yield and reduction of hydrogen sulfide production in the anaerobic digestion of waste activated sludge. Bioresource Technol. 243: 914–921. https://doi.org/10.1016/j.biortech.2017.07.036.

Duarte, M. S., Silva, S. A., Salvador, A. F., Cavaleiro, A. J., Stams, A. J. M., Alves, M. M. et al. (2018). Insight into the role of facultative bacteria stimulated by microaeration in continuous bioreactors converting LCFA to methane. Environ. Sci. Technol. 52(11): 6497–6507. https://doi.org/10.1021/acs.est.8b00894.

European Biogas Association. (2024). About biogas and biomethane. https://www.europeanbiogas.eu/about-biogas-and-biomethane/ (accessed 03 July 2024).

Fdz.-Polanco, M., Díaz, I., Pérez, S. I., Lopes, A. C. and Fdz.-Polanco, F. (2009). Hydrogen sulphide removal in the anaerobic digestion of sludge by micro-aerobic processes: pilot plant experience. Water Sci. Technol. 60(12): 3045–3050. https://doi.org/10.2166/wst.2009.738.

Fondriest Environmental, Inc. (2013). Dissolved Oxygen. In: Fundamentals of Environmental Measurements. https://www.fondriest.com/environmental-measurements/parameters/water-quality/dissolved-oxygen/ (accessed 03 July 2024).

Fu, H., Yuan, J. and Gao, H. (2015). Microbial oxidative stress response: Novel insights from environmental facultative anaerobic bacteria. Arch. Biochem. Biophys. 584: 28–35. https://doi.org/10.1016/j.abb.2015.08.012.

Girotto, F., Peng, W., Rafieenia, R. and Cossu, R. (2018). Effect of aeration applied during different phases of anaerobic digestion. Waste Biomass Valori. 9: 161–74. https://doi.org/10.1007/s12649-016-9785-9.

Hasegawa, S., Shiota, N., Katsura, K. and Akashi, A. (2000). Solubilization of organic sludge by thermophilic aerobic bacteria as a pretreatment for anaerobic digestion. Water Sci. Technol. 41(3): 163–169. https://doi.org/10.2166/wst.2000.0068.

Horne, A. J. and Lessner, D. J. (2013). Assessment of the oxidant tolerance of *Methanosarcina acetivorans*. FEMS Microbiol. Lett. 343(1): 13–19. https://doi.org/10.1111/1574-6968.12115.

Jagadabhi, P. S., Kaparaju, P. and Rintala, J. (2010). Effect of micro-aeration and leachate replacement on COD solubilization and VFA production during mono-digestion of grass-silage in one-stage leach-bed reactors. Bioresource Technol. 101(8): 2818–2824. https://doi.org/10.1016/j.biortech.2009.10.083.

Jasso-Chávez, R., Santiago-Martínez, M. G., Lira-Silva, E., Pineda, E., Zepeda-Rodríguez, A., Belmont-Díaz, J. et al. (2015). Air-adapted *Methanosarcina acetivorans* shows high methane production and develops resistance against oxygen stress. PLoS One 10: 121. https://doi.org/10.1371/journal.pone.0117331.

Jeníček, P., Horejš, J., Pokorná-Krayzelová, L., Bindzar, J. and Bartáček, J. (2017). Simple biogas desulfurization by microaeration – Full scale experience. Anaerobe 46: 41–45. https://doi.org/10.1016/j.anaerobe.2017.01.002.

Johansen, J.-E. and Bakke, R. (2006). Enhancing hydrolysis with microaeration. Water Sci. Technol. 53(8): 43–50. https://doi.org/10.2166/wst.2006.234.

Khademian, M. and Imlay, J. A. (2020). Do reactive oxygen species or does oxygen itself confer obligate anaerobiosis? The case of *Bacteroides thetaiotaomicron*. Mol. Microbiol. 114(2): 333–347. https://doi.org/10.1111/mmi.14516.

Khademian, M. and Imlay, J. A. (2021). How microbes evolved to tolerate oxygen. Trends Microbiol. 29(5): 428–440. https://doi.org/10.1016/j.tim.2020.10.001.

Khanal, S. K. and Huang, J.-C. (2006). Online oxygen control for sulfide oxidation in anaerobic treatment of high sulfate wastewater. Water Environ. Res. 78: 397–408. https://doi.org/10.2175/106143006x98804.

Krayzelova, L., Bartacek, J., Díaz, I., Jeison, D., Volcke, E. I. P. and Jenicek, P. (2015). Microaeration for hydrogen sulfide removal during anaerobic treatment: a review. Rev. Environ. Sci. Bio. 14(4): 703–725. https://doi.org/10.1007/s11157-015-9386-2.

Li, J., Ran, X., Zhou, M., Wang, K., Wang, H. and Wang, Y. (2022). Oxidative stress and antioxidant mechanisms of obligate anaerobes involved in biological waste treatment processes: A review. Sci. Total Environ. 838: 156454. https://doi.org/10.1016/j.scitotenv.2022.156454.

Lim, J. W. and Wang, J.-Y. (2013). Enhanced hydrolysis and methane yield by applying microaeration pretreatment to the anaerobic co-digestion of brown water and food waste. Waste Manage. 33(4): 813–819. https://doi.org/10.1016/j.wasman.2012.11.013.

Lim, J. W., Chiam, J. A. and Wang, J.-Y. (2014). Microbial community structure reveals how microaeration improves fermentation during anaerobic co-digestion of brown water and food waste. Bioresource Technol. 171: 132–138. https://doi.org/10.1016/j.biortech.2014.08.050.

Liu, C.-G., Xue, C., Lin, Y.-H. and Bai, F.-W. (2013). Redox potential control and applications in microaerobic and anaerobic fermentations. Biotechnol. Adv. 31: 257–65. http://dx.doi.org/10.1016/j.biotechadv.2012.11.005.

Liu, C.-G., Qin, J.-C. and Lin, Y.-H. (2017). Fermentation and redox potential. pp. 23–42. *In*: Jozala, A. F. (ed.). Fermentation Processes. Vol 2., 2nd ed. IntechOpen, London, U.K. https://doi.org/10.5772/64640.

Lu, Z. and Imlay, J. A. (2021). When anaerobes encounter oxygen: mechanisms of oxygen toxicity, tolerance and defence. Nat. Rev. Microbiol. 19(12): 774–785. https://doi.org/10.1038/s41579-021-00583-y.

Lyu, Z. and Lu, Y. (2018). Metabolic shift at the class level sheds light on adaptation of methanogens to oxidative environments. ISME J. 12(2): 411–423. https://doi.org/10.1038/ismej.2017.173.

Magalhães, C. P., Duarte, M. S., Pereira, M. A., Stams, A. J. M. and Cavaleiro, A. J. (2025). Facultative anaerobic bacteria enable syntrophic fatty acids degradation under micro-aerobic conditions. Biores. Technol. 417: 131829. https://doi.org/10.1016/j.biortech.2024.131829.

Magdalena, J. A., Angenent, L. T. and Usack, J. G. (2022). The measurement, application, and effect of oxygen in microbial fermentations: focusing on methane and carboxylate production. Fermentation 8(4): 138. https://doi.org/10.3390/fermentation8040138.

Montalvo, S., Huiliñir, C., Ojeda, F., Castillo, A., Lillo, L. and Guerrero, L. (2016). Microaerobic pretreatment of sewage sludge: Effect of air flow rate, pretreatment time and temperature on the aerobic process and methane generation. Int. Biodeter. Biodegr. 110: 1–7. https://doi.org/10.1016/j.ibiod.2016.01.010.

Montalvo, S., Vielma, S., Borja, R., Huiliñir, C. and Guerrero, L. (2018). Increase in biogas production in anaerobic sludge digestion by combining aerobic hydrolysis and addition of metallic wastes. Renew. Energ. 123: 541–548. https://doi.org/10.1016/j.renene.2018.02.004.

Morais, B. P., Magalhães, C. P., Martins, G., Pereira, M. A. and Cavaleiro, A. J. (2024). Effect of micro-aeration on syntrophic and methanogenic activity in anaerobic sludge. Appl. Microbiol. Biotechnol. 108: 192. https://doi.org/10.1007/s00253-023-12969-4.

Nguyen, D. and Khanal, S. K. (2018). A little breath of fresh air into an anaerobic system: How microaeration facilitates anaerobic digestion process. Biotechnol. Adv. 36(7): 1971–1983. https://doi.org/10.1016/j.biotechadv.2018.08.007.

Nguyen, D., Wu, Z., Shrestha, S., Lee, P.-H., Raskin, L. and Khanal, S. K. (2019). Intermittent micro-aeration: New strategy to control volatile fatty acid accumulation in high organic loading anaerobic digestion. Water Res. 166: 115080. https://doi.org/10.1016/j.watres.2019.115080.

Ramos, I., Pérez, R., Reinoso, M., Torio, R. and Fdz-Polanco, M. (2014). Microaerobic digestion of sewage sludge on an industrial-pilot scale: The efficiency of biogas desulphurisation under different configurations and the impact of O_2 on the microbial communities. Bioresource Technol. 164: 338–346. https://doi.org/10.1016/j.biortech.2014.04.109.

Stackebrandt, E. (2004). The phylogeny and classification of anaerobic bacteria. pp. 1–25. *In*: Nakano, M. M. and Zuber, P. (eds.). Strict and Facultative Anaerobes. Medical and Environmental Aspects. Horizon Bioscience, Wymondham, Norfolk, U.K.

Schoeffler, M., Gaudin, A.-L., Ramel, F., Valette, O., Denis, Y., Hania, W. B. et al. (2019). Growth of an anaerobic sulfate-reducing bacterium sustained by oxygen respiratory energy conservation after O_2-driven experimental evolution. Environ. Microbiol. 21: 360–373. https://doi.org/10.1111/1462-2920.14466.

Thauer, R. K. (2019). Methyl (alkyl)-coenzyme M reductases: nickel F-430-containing enzymes involved in anaerobic methane formation and in anaerobic oxidation of methane or of short chain alkanes. Biochemistry-US 58: 5198–5220. https://doi.org/10.1021/acs.biochem.9b00164.

Tsapekos, P., Kougias, P. G., Vasileiou, S. A., Lyberatos, G. and Angelidaki, I. (2017). Effect of micro-aeration and inoculum type on the biodegradation of lignocellulosic substrate. Bioresource Technol. 225: 246–253. https://doi.org/10.1016/j.biortech.2016.11.081.

Wu, C., Zhou, Y., Wang, P. and Guo, S. (2015). Improving hydrolysis acidification by limited aeration in the pretreatment of petrochemical wastewater. Bioresource Technol. 194: 256–262. https://doi.org/10.1016/j.biortech.2015.06.072.

Wu, C., Li, Y., Zhou, Y., Li, Z., Zhang, S. and Liu, H. (2018). Upgrading the Chinese biggest petrochemical wastewater treatment plant: Technologies research and full scale application. Sci. Total Environ. 633: 189–197. https://doi.org/10.1016/j.scitotenv.2018.03.164.

Wu, Z., Nguyen, D., Lam, T. Y. C., Zhuang, H., Shrestha, S., Raskin, L. et al. (2021). Synergistic association between cytochrome bd-encoded *Proteiniphilum* and reactive oxygen species (ROS)-scavenging methanogens in microaerobic-anaerobic digestion of lignocellulosic biomass. Water Res. 190: 116721. https://doi.org/10.1016/j.watres.2020.116721.

Xu, S., Selvam, A. and Wong, J. W. C. (2014). Optimization of micro-aeration intensity in acidogenic reactor of a two-phase anaerobic digester treating food waste. Waste Manage. 34(2): 363–369. https://doi.org/10.1016/j.wasman.2013.10.038.

Xu, H., Li, Y., Hua, D., Zhao, Y., Chen, L., Zhou, L. et al. (2021). Effect of microaerobic microbial pretreatment on anaerobic digestion of a lignocellulosic substrate under controlled pH conditions. Bioresource Technol. 328: 124852. https://doi.org/10.1016/j.biortech.2021.124852.

Yang, H., Deng, L., Wu, J., Wang, W., Zheng, D., Wang, Z. et al. (2021). Intermittent air mixing system for anaerobic digestion of animal wastewater: operating conditions and full-scale validation. Bioresource Technol. 335: 125304. https://doi.org/10.1016/j.biortech.2021.125304.

Zabranska, J. and Pokorna, D. (2018). Bioconversion of carbon dioxide to methane using hydrogen and hydrogenotrophic methanogens. Biotechnol. Adv. 36: 707–720. https://doi.org/10.1016/j.biotechadv.2017.12.003.

Zhang, Z., Yu, Y., Xi, H. and Zhou, Y. (2021). Review of micro-aeration hydrolysis acidification for the pretreatment of toxic and refractory organic wastewater. J. Clean. Prod. 317: 128343. https://doi.org/10.1016/j.jclepro.2021.128343.

Zhen, X., Zhang, X., Li, S., Li, M. and Kang, J. (2020). Effect of micro-oxygen pretreatment on gas production characteristics of anaerobic digestion of kitchen waste. J. Mater. Cycles Waste 22(6): 1852–1858. https://doi.org/10.1007/s10163-020-01072-9.

Zhu, M., Lü, F., Hao, L.-P., He, P.-J. and Shao, L.-M. (2009). Regulating the hydrolysis of organic wastes by micro-aeration and effluent recirculation. Waste Manage. 29: 2042–2050. https://doi.org/10.1016/j.wasman.2008.12.023.

Zitomer, D. H. and Shrout, J. D. (1998). Feasibility and benefits of methanogenesis under oxygen-limited conditions. Waste Manage. 18: 107–116. https://doi.org/10.1016/S0956-053X(98)00008-7.

Chapter 8

Microbial Insights from Metagenomics

Anaerobic Communities Explored

Mehak Manchanda,[1] *Palak Gupta,*[1] *Arathi Radhakrishnan,*[1] *Abhishek Chauhan,*[2] *Raj Kishor Kapardar*[3] and *Rajpal Srivastav*[1,*]

8.1 Introduction

Microbes function as a crucial entity in the ecological homeostasis. Be it human microbiome, sludge digesters, fermenters, pharmacy, or the biogeochemical cycles all require bacteria to function and produce the desired yield or product (Anhê et al. 2015, Radhakrishnan et al. 2023). In the past decades, through various culturing techniques microbes are made to grow, their genomes being entirely decoded, for further research under the branch of science as genomics. Interestingly, more than 95% of the microbes in the environment cannot be cultivated (Handelsman et al. 1998, Chauhan et al. 2017), making it one of the biggest roadblocks in research and development. Metagenomics has emerged as a solution to this problem (Browne et al. 2016). Instead of studying individual genome of the species, cultivated in laboratories, genomes of an entire community are being studied at a time, directly from its natural environment, this is known as metagenomics.

Moreover, laboratory culturing can be time laborious as it relies on the ability to culture the bacterial species of interest. Further, despite the significant advancements in

[1] Amity Institute of Biotechnology, Amity University Uttar Pradesh, Noida, India.
[2] Amity Institute of Environmental Toxicology, Safety and Management, Amity University Uttar Pradesh, Noida, India.
[3] The Energy and Resources Institute, New Delhi, India.
* Corresponding author: rajpalsrivastav@gmail.com, rsrivastav2@amity.edu

DOI: 10.1201/9781003506898-8

metagenomics and its potential to shape the future, it is crucial to thoroughly evaluate the data collected from it by integrating it with other clinical information. This is necessary because microbes identified through metagenomics may exist in a dormant stage especially in clinical settings. In general, a sequential technical procedure is followed in metagenomics which starts at the sample collection, followed by genome extraction, metagenomic library preparation, sequencing, and analyzing the data using bioinformatics tools followed by *in-silico* or functional analysis. The last step is demarcated by functional profiling, this involves predicting the metabolic capabilities of different microorganisms, including their roles in anaerobic processes like fermentation, methanogenesis, or sulphate reduction. Various microbes have been explored with metagenomic applications utilizing high-throughput Next-Generation Sequencing technologies (NGS) that have provided the opportunity to explore microbes without the requirement of culturing. Various advanced sequencing technologies are now being employed to obtain significant genomic analysis, these include next generation sequencing technologies and their hybrids platforms. In recent times, metagenomics has played significant roles in exploring the composition of microbial communities and the metabolic pathways present in various environments especially anaerobic microorganisms (Srivastav et al. 2013, Willeghems et al. 2016). This approach, which focuses on the genomes of microorganisms, offers a valuable means to comprehend the diverse range of functions and metabolic activities occurring in anaerobic environment (Gophna et al. 2017, Srivastav and Suneja 2019). Various aspects associated with this has been discussed in the following sections (Table 8.1).

Table 8.1. Representation of examples of few microbes explored utilizing metagenomic approaches.

SN	Anaerobes Explored	Site (s) location/ Source	Method(s)/approach(es)	References
1	*F. nucleatum*	Faecal samples, Oral squamous cell, Colorectal Adenomas	16S rRNA amplicon, Metagenomic Sequencing	Xie et al, 2021
2	*Chlorobium, Desulphobacter, Mycolicibacterium*	Sediment, biomass, Lake Geneva in Switzerland	Shotgun sequencing, Illumina NovaSeq6000	Suarez et al. 2022 Gophna et al. 2017 Bashiri et al. 2022
3	*Syntrophomonas*	Anaerobic digesters	16S rDNA, metagenome sequencing	Meng et al. 2022
4	*Camphylobacter*	Faecal samples infected chickens	Nanopore sequencing	Bloemen et al. 2023
5	*Akkermansia muciniphila*	Human gut	NGS, 16s RNA	Zhou et al. 2017
6	*A. muciniphilia*	Mice gut	HiFi and NGS	Rodrigues et al. 2022
7	*Ca. Scalindua Ca. Brocadiales*	Subsea floor tunnel, Norway	NGS, 16s RNA	Suarez et al. 2022 Gophna et al. 2017
8	*Clostridium difficile*	faecal samples	Illumina HiSeq	Sakai et al. 2022
9	*Klebsiella*	Gallstones	16S rRNA profiling, Illumina HiSeq. 2500 platform	Kose et al. 2018

8.2 Anaerobic Microbes and Metagenomics

Metagenomics has further made analysis of all genomic data easier as it not only names marker genes but the intact set, therefore fostering taxonomy analysis. In clinicals metagenomics and next generation sequencing (mNGS) is already being tapped to detect pathogens and it has applications in microbe interactions analysis (Srivastav et al. 2013). Soil has a vast array of microorganisms, and research on metagenomics has revealed that soil can hold a wealth of potential antibiotics and antifungal drugs.

The soil metagenomic library was used to evaluate and name the broad-spectrum antibacterial antibiotics Turbomycin A and Turbomycin B (Gillespie et al. 2002). Marine microorganisms have more potential than soil microbes. Over 70% of the Earth's surface is made up of rivers, lakes, but instead oceans, which result in a wide range of species, genetic makeup, biological traits among the microorganisms living in water. Specialized metabolites and metabolic regulatory systems are present in marine bacteria. Various marine secondary metabolites' synthesis routes are studied using metagenomic approaches. One such instance is the polyketone biosynthetic gene cluster associated with marine actinomycetes-related compounds, which could provide as a source of novel pharmaceuticals. Most river and lake systems have significant concentrations of Antibiotic-Resistant Bacteria (ARBs), namely Methicillin-Resistant *Staphylococcus Aureus* (MRSA) and Vancomycin-Resistant *Enterococcus* (VRE) (Nnadozie and Odume 2019).

In a recent work, at the Lake Cuochuolong, an alkaline saline lake located on the Tibetan Plateau, metagenomes were compiled with 8 different assembly programs. A whole of 17 phyla were identified, each with a comparative quantity bigger than 0.1%. Among them, Alphaproteobacteria was the most abundant group in the collected section. Other notable phyla included Ciliophora (44.05%) (Tao et al. 2023). The success of this study can be in addition to this aspect alone, several portable sequencing technologies like MinION have facilitated accurate identification of *Campylobacter* in caecal samples obtained from infected chickens within a timeframe of less than five hours. This contrasts with the conventional methods that typically require more than three days, attributed to the utilization of metagenomics (Fig. 8.1).

Current applied long-read metagenomics with nanopore sequencing are exploited to analyze HQ MAGs (High Quality Metagenome-Assembled genome) of a faecal sample to explore the gut microbiome. Diabetes, obesity, metabolic malfunctions, and stoma are in the spotlight through metagenomic studies of certain anaerobes which are showing promising outputs as their cure. Several illnesses, including cancers, autoimmune disorders, and non-alcoholic fatty liver disease, kidney disorders may be linked to the gut microbiota (Belzer and de Vos 2012, Radhakrishanan et al. 2022). Metagenomics may be used to track infectious disease epidemics in hospitals and communities, as well as finding drug-resistant genes for pathogens (Brahe et al. 2015, Srivastav et al. 2019). By using mNGS testing, five species were found in the cerebral fluid: *Neisseria meningitidis, Streptococcus agalactiae, Candida albicans, Mycobacterium fortuitum,* and *Mycobacterium abscessus* (Miller et al. 2019).

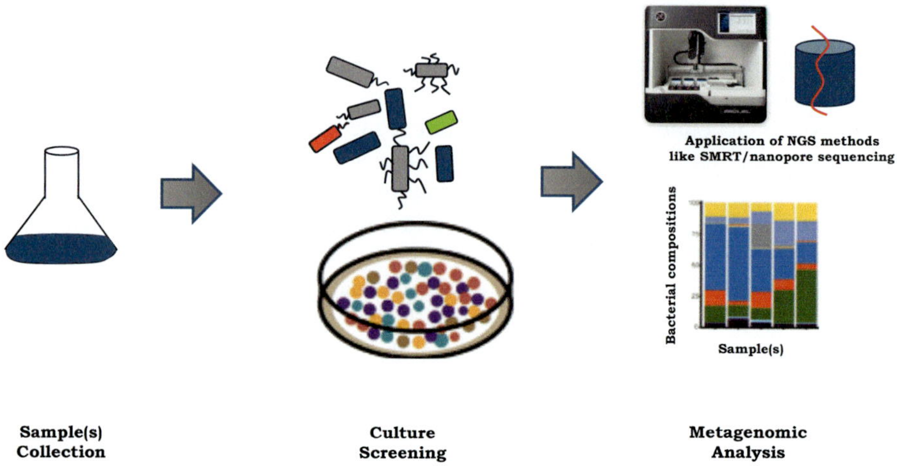

Figure 8.1. Overview of the metagenomic analysis in culture enriched approach.

8.2.1 Oral Carcinoma

16s rRNA gene sequencing and metagenomics have revealed different types of bacteria are found in the oral cavity of humans (Deo et al. 2019). A sizeable part of the oral bacteriome is made up of anaerobes, which are crucial to the development of multi-species biofilms affixed to a variety of anatomical locations. The most common kind of head and neck cancer subgroup, Oral Squamous Cell Carcinoma (OSCC), is a leading global source of morbidity and mortality (Johnson et al. 2011).

In some communities, especially in Southern Asia and the Pacific islands, OSCC has an exceptionally high frequency and is the key reason of cancer-related death, for males in India and Sri Lanka. Oral squamous cell carcinoma (OSCC) and other pathologic lesions of the oral cavity are linked to bacterial biofilms, and specific oral taxa may be found inside tumours, including deep biopsy samples. *Fusobacterium nucleatum* is one microbe that plays a significant role in the growth and maturation of biofilms that follow dysbiotic alterations in dental plaque. Using NGS or 16s rRNA sequencing have been used to validate the fact that *F. nucleatum* is higher in OSCC than in normal tissue (McIlvanna et al, 2021). In addition to 16S rRNA amplicon sequencing, metagenomic sequencing can be conducted using the same DNA. Metagenomic data showed a higher richness compared to the 16S rRNA data, indicating that it identified more taxa. Metagenomic sequencing examines all the genomic DNA in a sample, providing a broad representation of microbial genomes, while 16S rRNA sequencing targets and reads a specific region of the 16S gene, providing less comprehensive information (McIlvanna et al. 2021). The choice between the two methods depends on factors such as the sample type, desired analyses, taxonomy resolution, and available bioinformatic capabilities (Pereira et al. 2020, Suneja and Srivastav 2021).

8.2.2 Colorectal Diseases

In case-regulated studies of colorectal adenomas utilizing 16S rRNA gene scans, with one observing fusobacterial enrichments in their patient unit. A total of 16 faecal samples were collected from 16 patients who were hospitalized. These patients were then categorized into three groups: Group A (adenoma), Group C (colorectal cancer), or Group H (haemorrhoids). The DNA was extracted followed by 16S rRNA gene amplicon and SMS (Shallow Metagenomic) analyses. Furthermore, comprehensive data collection involves gathering patient information regarding their smoking and alcohol intake history, along with their history of antibiotic use. V3-V4 regions of the 16S rRNA gene sequences was amplified. The amplified PCR fragments were then purified and combined. Finally, the sequencing of the mixed PCR fragments was carried out for the analysis of the 16S rRNA amplicon data. The sequences were subsequently categorized into Operational Taxonomic Units (OTUs) using a 97% identity threshold. The Usearch beta_div tool was employed to assess the beta-diversity distance and disparities between the samples and the OTU. Notably, Group A had 49.76% Bacteroidota, compared to Groups C and H, which had 61.66 and 75.79%, respectively. Furthermore, the SMS data revealed that the fraction of Bacteroidota in Groups C and H was higher than the 16S rRNA data indicated (Kostic et al. 2013).

8.2.3 Lipid Degradation

Domestic wastewater treatment is considered after lipid degradation but at low temperature the lipases secreted by psychrophiles failed to do so. Hence the research used metagenomics along with the combination of meta proteomics to point at the anaerobes to enhance lipid degradation in domestic wastewater treatment (Affes et al. 2017).

The inadequate breakdown of lipids under low temperature conditions may be attributed to various factors, including insufficient production and activity of lipase enzymes, sluggish transportation of fatty acids, and a slow beta-oxidation pathway (Black and DiRusso 2003). Consequently, the objective of the current investigation is to explore the lipolytic capabilities of psychrophilic bacteria that have adapted to anaerobically treat domestic wastewater at temperatures of 4 and 15°C. Metagenomics is employed to identify potential lipase genes that are actively expressed, as well as other lipolytic biomarkers like fadL genes (Bashiri et al. 2022). Out of the 40 Metagenome-Assembled Genomes (MAGs) that were identified as potential lipolytic organisms, only three of them (*Chlorobium, Desulfobacter*, and *Mycolicibacterium*) were found to be both common and abundant in all the reactors. These three MAGs had a relative abundance of at least 1% in each reactor. To investigate the performance of anaerobic membrane bioreactors (AnMBRs) under different temperature conditions, four AnMBRs with a working volume of 1L were developed. The utilization of cold-adapted biomass allowed these reactors to function at temperatures of 4 and 15°C. The decision to select these specific temperatures was influenced by a prior investigation to immunize the reactors, potential psychrophilic biomass was gathered from the sediment and soils of Lake Geneva in Switzerland

and Svalbard in Norway. These locations experience annual minimum and maximum temperatures of –11 and 21°C for Lake Geneva, and –16 and 6°C for Svalbard, respectively. The primary domestic influent used as feed for the reactors was obtained from an activated sludge plant located in Tudhoe Mill, County Durham, UK. The feed was introduced into the reactors in two different forms: Sterile (treated with ultraviolet light to exclude mesophilic biomass) or non-sterile (Bashiri et al. 2022).

To ensure successful acclimation of the psychrophilic biomass to the operating temperatures (4 and 15°C) and feed types, a succession of batch and continuous experiments were conducted over a period of more than three years, as described in their studies. In this study, the reactors were re-acclimated for two months, with a hydraulic retention time of 60 hours. The entire metagenome analysis revealed that the most annotated genes were β-glucosidase, β-hexosaminidase, cellulase, α-amylase, α-galactosidase, and endo-β-xylanase. This significant difference in gene annotation suggests that cells may have different regulatory systems for sugar-metabolizing genes. There exists a mechanism called Carbon Catabolite Repression (CCR), which inhibits the expression of genes that allow cells to utilize secondary carbon sources when easily accessible sugars are present like catabolite repression resistance gene phosphotransferase system sugar specific EII component (PTS-EII) (Bashiri et al. 2022).

The CCR regulatory system selects the most easily utilizable energy source. In the case of lipases, CCR suppresses their production when glucose is abundant. Furthermore, the expression of proteases also influences lipase production. For example, in *Bacillus subtillis*, the accumulation of amino acids triggers the cells to produce more proteases and lipase production get reduced (Razzaq et al. 2019). The association of transporters is important in deducing these mechanisms (Srivastav et al. 2014, López and Bogaert 2021).

8.2.4 Butyrate Degradation

Anaerobic Digestion (AD) has gained widespread recognition as an efficient solution for processing and reusing solid waste, including straws, food waste, and sewage, to produce renewable resources (Pilarska et al. 2023). The AD process consists of four main stages: hydrolysis, acidogenesis, acetogenesis, and methanogenesis (Lim et al. 2020). Short-chain fatty acids, particularly butyrate, play a crucial role as major metabolic intermediates connecting acidogenesis and methanogenesis. However, due to the unfavourable energetics of reactions, butyrate tends to accumulate at high organic loading rates, leading to acidification and reduced gas production (Besten et al. 2013). Higher concentration may have inhibitory effects (Li et al. 2017). Therefore, the degradation of butyrate is of utmost importance for the overall success of the anaerobic process. The symbiotic bacteria could potentially enhance the efficiency of anaerobic digestion (AD). Initially, the Synergistic Butyrate-Oxidizing Bacteria (SBOB) are carefully selected. Syntrophomonas was the first documented bacterium that exhibited syntrophic behaviour by degrading butyrate in co-cultures with methanogens or sulphate-reducing prokaryotes (Dyksma et al. 2022). Subsequently, several other genera have been identified, including Syntrophococcus,

Syntrophus, Syntrophospora, Synthrophothermus, and Thermosyntropha (Li et al. 2022).

To assess the effectiveness of the anaerobic digestion system, various gas and liquid parameters were closely monitored, including methane concentration, volumetric biogas generation, levels of Volatile Fatty Acids (VFAs), alkalinity, and pH. High-throughput sequencing of 16S rRNA were performed on the illumina platform to examine the fluctuations in microbial diversity analysis. Moreover, sequencing was carried out for variations in metabolomics. By integrating the findings from these analyses, the group was able to gain valuable insights into the synergistic communities in a complex environment, identify influential factors, and elucidate crucial metabolic pathways. This comprehensive understanding could potentially contribute to the development of strategies aimed at mitigating acid accumulation. In another research, Syntrophomonas is found to be the butyrate oxidizing bacteria adhering beta-oxidation in a microbial consortium while acetate was oxidized by Synergistaceae and Mesotoga. This revelation was made using the amalgamation of the 16S rDNA and metagenome sequencing investigations (Meng et al. 2022).

Anaerobic sludge digesters are enhanced by the presence of certain methanogens depending upon the feeding factors found methanosarcina worked actively at high OLR (Organic Loading Rate) followed by Mesotoga in different reactors as experimented in 2023, MAGs were analyzed and hence domains located with same gene patterns of butyrate degradation (Kong et al. 2019).

8.2.5 Anti-Microbial Resistance

Antibiotics resistance is a serious concern specially the emergence of multi-drug resistance in microorganisms. There are various mechanisms which can be used by the microbes for avoiding cellular stress generated by the anti-microbials (Madhavi et al. 2017). So, there is constant need for re-developing and re-discovering the strategies at molecular levels (Radhakrishnan et al. 2024). With the help of metagenomics, AMR analysis can be performed. In an analysis, the caecal samples collected from infected chickens were effectively examined using a method that detected Campylobacter in under five hours, a significant advancement from traditional techniques that required over 4–5 days. An amplicon sequencing method that can accurately identify quarantine plant pathogens with increased sensitivity. *In-situ* nanopore amplicon sequencing was utilized to evaluate biodiversity. To streamline nanopore sequencing, the BQV method was created for on-site DNA extraction and library preparation (Beatriz et al. 2021). When paired with the portable VolTRAX device, this optimized protocol enables the production of high-quality nanopore libraries. The final BQV protocol, utilizing portable devices for cell lysis, DNA purification, and library preparation, allows for on-site mNGS library creation from complex samples in a short time. The libraries generated using the BQV method demonstrate a comparable throughput in shorter time (Bloemen et al. 2023).

The Claremont Bio DNAexpress kit was utilized to create an on-site DNA extraction procedure, which involved homogenizing and lysing samples in a battery-powered Omnilyse X bead-beating tube. The original protocol was adjusted to enhance DNA purity, but initially it produced subpar DNA results. This protocol yielded comparable amounts of high purity DNA as the current laboratory-based EQ method, albeit with shorter fragment sizes. It is important to note that the taxonomic classification of the sample is influenced by the sequencing workflow. Such method improvements are useful for enhancing the output of metagenomic analysis (Ghatak et al. 2013).

8.2.6 Association with Metabolic Disorders

The occurrence of metabolic disorders is impacted by the disruption of homeostasis and the maintenance of the intestinal barrier. The primary function of the intestinal mucosal barrier is to strike a balance between absorbing necessary nutrients and preventing the movement of harmful pathogens (Di Tommaso et al. 2021).

This barrier is reinforced by Tight Junctions (TJs), Adherens Junctions (AJs), and desmosome complexes within the intestinal epithelium. Probiotics and their derivatives, including microvesicles, can boost the expression of these junctions, thereby safeguarding the integrity of the intestinal epithelium (Fortea et al. 2021).

When the integrity of the intestinal mucosa is compromised, it can result in the onset of bowel diseases, Ulcerative Colitis (UC), Crohn's Disease (CD), etc. These inflammatory conditions include excessive immune response against microbiome (Ramos et al. 2019). Data indicates that the use of antibiotics in an early age can modify the microbes and heighten the susceptibility to gut problems where intestinal homeostasis is disturbed. Furthermore, studies have pinpointed variations in the gut microbiota composition between healthy individuals and those with IBD. Specifically, individuals with UC exhibit notably lower levels of *Akkermansia muciniphila* (*A. muciniphila*) (Patangia et al. 2022). It has also been noted that *A. muciniphila* or a specific outer membrane protein known as Amuc_1100 can alleviate colitis triggered by Dextran Sulphate Sodium (DSS) in a mouse model (Zheng et al. 2023). The results may have variations as per the role of complexities including mutilgenic associations, which need to be deciphered as role of organelles communication also determines overall phenotype and its outcome (Grover et al. 2019).

The involvement of *A. muciniphila* in the resistance against diabetes has sparked debate, but there is evidence indicating its potential role in the development of metabolic and autoimmune diseases. Studies have shown that mice fed on a high-fat diet for 16 weeks experienced increased inflammation markers in their gut microbiota, as well as hyperinsulinemia, hyperglycemia, and higher levels of serum leptin. Interestingly, the abundance of *A. muciniphila* decreased after 3 weeks of the diet, preceding the development of peripheral insulin resistance and type 2 diabetes (Rodrigues et al. 2022). Refractory type 2 diabetes, which refers to the inability to achieve optimal glycemic control, has also been linked to a lower relative abundance of *A. muciniphila* compared to diabetic individuals. This was accompanied by improvements in the liver function, reduced levels of pro-inflammatory factors in the blood, as well as a decrease in gluco/lipotoxicity and oxidative stress. When it comes to type 1 diabetes, the cytokine IFN-gamma plays a significant role in the onset of the disease. Mice lacking IFN-gamma have shown better glucose tolerance and increased abundance of *A. muciniphila* in their gut microbiota compared to wild-type mice (Shih et al. 2020). On the other hand, IFN-gamma-deficient mice without *A. muciniphila* did not show any improvement in glucose tolerance. Additionally, a study found that NOD mice with a low incidence of type 1 diabetes had a higher abundance of *A. muciniphila* in their gut microbiota. Moreover, the administration of *A. muciniphila* to a colony of NOD mice with a high incidence of type 1 diabetes delayed the onset of the disease (Greer et al. 2016).

Thus, in conclusion, an imbalance in the gut microbiota has been suggested as a risk factor for the development of metabolic diseases.

8.2.7 Ammonium Oxidation

Anammox bacteria, which are strict anaerobes, can thrive in environments with limited oxygen availability, such as dense biofilms and areas with low oxygen concentrations in the marine ecosystem. These bacteria have also been discovered in deep subseafloor sediments. The biofilms in this region are characterized by the presence of manganese oxide and iron hydroxide biominerals, and the redox reactions occurring within these biofilms can lead to acidification (Suarez et al. 2022). Through 16S rRNA gene sequencing, it has been found that the microbial community in this area consists of a significant number of microorganisms involved in nitrogen transformation, such as ammonium-oxidizing bacteria and archaea. The data also indicated the presence of marine anammox bacteria known as Candidatus Scalindua, which were classified as amplicon sequence variants. In some samples, a group of unclassified bacteria known as Ca. Brocadiales were found to be the second most abundant taxa. However, it is still uncertain whether these unclassified sequences belong to anammox bacteria. To further investigate the Ca. Brocadiales community and gain a better understanding of their taxonomy and potential functions, shotgun metagenomic sequencing was employed.

A study focusing on Metagenome-Assembled Genomes (MAGs) from two tunnel sites identified seven Oslofjord Tunnel MAGs (OFTMs) classified within the order Ca. Brocadiales. Among these, three MAGs (OFTM134, OFTM214, and OFTM301) were present in five out of eight samples, with their relative abundance ranging from 0.05 to 5.37%. The remaining four MAGs were affiliated with the genus Ca. Scalindua (Somerville et al. 2019). The unclassified MAGs may represent previously unknown anammox taxa. These bacteria were identified in metagenomes extracted from a biofilm situated in the subseafloor. To indicate their expected ability for the anammox process, suggested names are Ca. Anammoxibacter for the newly recognized genus and Ca. Anammoxibacteraceae for the family. This research makes a significant contribution in analysis of microbial diversity in oceans offering possibilities for innovative biotechnological applications (Yulandi et al. 2020).

The study of metagenomes is aiding in the research of bioaugmented Zero Water Exchange Aquaculture Production Systems (ZWEAPS) that have minimal or no water exchange. These systems, which include the anaerobic ammonia oxidizing bacterium Candidatus Kuenenia, do not experience ammonia accumulation. This research is being conducted by Nair RR, Rangaswamy B, Sarojini BSI, and Joseph V (Nair et al. 2020).

8.2.8 Human Gut Health

The genomic-based method can be used for selectively cultivating specific bacterial phenotypes in the human intestinal microbiota. The findings showed a significant correlation between the two types of samples at the species level, with an average of 93% of raw reads being shared between the original faecal sample and the cultured bacterial community (Forster et al. 2019).

Another study has demonstrated gut microbiota can be cultured using a single growth medium from a faecal sample. However, to match the species detection sensitivity of metagenomic sequencing, more than 8×10^6 distinct colonies would need to be picked. The study proposed the hypothesis that sporulation, under appreciated phenomenon may have association with transmission of microbiota among humans (Browne et al. 2016).

Microbiota profiling and sequencing involved amplifying the full-length 16S rRNA gene via the Polymerase Chain Reaction (PCR) method. Capillary sequencing was then used to acquire the gene sequences, which were aligned in the Ribosomal Database Project (RDP) and categorized into operational taxonomic units (OTUs) (Church et al. 2020). The R package seqinr version 3.1 was utilized to evaluate OTU similarity, with a species-level threshold of 98.7%. Comparisons with the Human Microbiome Project (HMP) were made by comparing the 16S rRNA gene sequences of cultured bacteria to define a species, using a sequence similarity threshold of 97%. The extracted genomic DNA was then sequenced on the Illumina HiSeq platform, with a subsequent assembly and annotation of the sequences for further analysis (Johnson et al. 2019).

8.3 Conclusive Remarks

Metagenomics is a highly useful technique to analyze the unculturable microorganisms employing the NGS methods including long read sequencing, Oxford Nanopore Technologies MinION sequencing, etc., for rapid identification and analysis of the sample. The application of mNGS has accelerated the pace of the study of anaerobes. More detailed data is required to verify the precision of metagenomics in determining microbial abundance in unexplored niches.

References

Affes, M., Aloui, F., Hadrich, F., Loukil, S. and Sayadi, S. (2017). Effect of bacterial lipase on anaerobic co-digestion of slaughterhouse wastewater and grease in batch condition and continuous fixed-bed reactor. Lipids Health Dis. 2017 Oct 10; 16(1): 195. doi: 10.1186/s12944-017-0587-2. PMID: 29017503; PMCID: PMC5634824.

Anhê, F. F., Roy, D., Pilon, G., Dudonné, S., Matamoros, S., Varin, T. V. et al. (2015). A polyphenol-rich cranberry extract protects from diet-induced obesity, insulin resistance and intestinal inflammation in association with increased Akkermansia spp. population in the gut microbiota of mice. Gut. 2015 Jun; 64(6): 872–83. doi: 10.1136/gutjnl-2014-307142. PMID: 25080446.

Arumugam, K., Bessarab, I., Haryono, M. A. S., Liu, X., Zuniga–Montanez, R. E., Roy, S. et al. (2021). Recovery of complete genomes and non-chromosomal replicons from activated sludge enrichment microbial communities with long read metagenome sequencing. Npj Biofilms Microbiomes 7: 23. doi: 10.1038/s41522-021-00196-6.

Badri, D. V., Jackson, M. I. and Jewell, D. E. (2021). Dietary protein and carbohydrate levels affect the gut microbiota and clinical assessment in healthy adult cats. J. Nutr 151: 3637–3650. doi: 10.1093/jn/nxab308.

Barry, K. A., Middelbos, I. S., Vester Boler, B. M., Dowd, S. E., Suchodolski, J. S., Henrissat, B. et al. (2012). Effects of dietary fiber on the feline gastrointestinal metagenome. J. Proteome Res. 11: 5924–5933. doi: 10.1021/pr3006809.

Baruch, E. N., Youngster, I., Ben-Betzalel, G., Ortenberg, R., Lahat, A., Katz, L. et al. (2021). Fecal microbiota transplant promotes response in immunotherapy-refractory melanoma patients. Science 371: 602–609. doi: 10.1126/science.abb5920 - DOI - PubMed.

Bashiri, R., Allen, B., Shamurad, B., Pabst, M., Curtis, T. P. and Ofiţeru, I. D. (2022). Looking for lipases and lipolytic organisms in low-temperature anaerobic reactors treating domestic wastewater. Water Res. 2022 Apr 1; 212: 118115. doi: 10.1016/j.watres.2022.118115. Epub 2022 Jan 22. PMID: 35092910.

Beatriz, Ferrari, R. G., Panzenhagen, P., Carolina, A. and Conte-Junior, C. A. (2021). Antimicrobial resistance gene detection methods for bacteria in animal-based foods: a brief review of highlights and advantages. Microorganisms 9(5): 923–923. https://doi.org/10.3390/microorganisms9050923.

Belzer, C. and de Vos, W. M. (2012). Microbes inside-from diversity to function: The case of Akkermansia. ISME Journal 6(8): 1449–1458.

Bibbò, S., Dore, M. P., Pes, G. M., Delitala, G. and Delitala, A. P. (2017). Is there a role for gut microbiota in type 1 diabetes pathogenesis? Ann. Med. 49: 11–22. doi: 10.1080/07853890.2016.1222449.

Black, P. N. and DiRusso, C. C. (2003). Transmembrane movement of exogenous long-chain fatty acids: proteins, enzymes, and vectorial esterification. Microbiol. Mol. Biol. Rev. 67(3): 454–472.

Bloemen, B., Gand, M., Vanneste, K., Marchal, K., Roosens, N. H. C. and De Keersmaecker, S. C. J. (2023). Development of a portable on-site applicable metagenomic data generation workflow for enhanced pathogen and antimicrobial resistance surveillance. Sci Rep. 2023 Nov 11; 13(1): 19656. doi: 10.1038/s41598-023-46771-z. PMID: 37952062; PMCID: PMC10640560.

Boe, K., Batstone, D. J., Steyer, J.-P. and Angelidaki, I. (2010). State indicators for monitoring the anaerobic digestion process. Water Res. 44: 5973–5980. doi: 10.1016/j.watres.2010.07.043.

Bolyen, E., Rideout, J. R., Dillon, M. R., Bokulich, N. A., Abnet, C. C., Al-Ghalith, G. A. et al. (2019). Reproducible, interactive, scalable and extensible microbiome data science using QIIME 2. Nat. Biotechnol. 37: 850–852. doi: 10.1038/s41587-019-0190-3 - DOI - PMC - PubMed.

Bonk, F., Popp, D., Weinrich, S., Sträuber, H., Kleinsteuber, S., Harms, H. et al. (2018). Intermittent fasting for microbes: how discontinuous feeding increases functional stability in anaerobic digestion. Biotechnol. Biofuels 11: 274. doi: 10.1186/s13068-018-1279-5.

Bradshaw, J. W. S., Goodwin, D., Legrand-Defrétin, V. and Nott, H. M. R. (1996). Food selection by the domestic cat, an obligate carnivore. Comp Biochem Physiol A Physiol 114: 205–209. doi: 10.1016/0300-9629(95)02133-7.

Brahe, L. K., Le Chatelier, E., Prifti, E., Pons, N., Kennedy, S., Hansen, T. and Larsen, L. H. (2015). Specific gut microbiota features and metabolic markers in postmenopausal women with obesity. Nutrition & Diabetes 5: e159. [PMC free article] [PubMed] [Google Scholar].

Brill, B., Amir, A. and Heller, R. (2019). Testing for differential abundance in compositional counts data, with application to microbiome studies. arXiv. preprint. arXiv:1904.08937.

Browne, H. P., Forster, S. C., Anonye, B. O., Kumar, N., Neville, B. A., Stares, M.D. et al. (2016). Culturing of 'unculturable' human microbiota reveals novel taxa and extensive sporulation. Nature. 2016 May 26; 533(7604): 543–546. doi: 10.1038/nature17645. Epub 2016 May 4. PMID: 27144353; PMCID: PMC4890681.

Cabral, L., Persinoti, G. F., Paixão, D. A. A., Martins, M. P., Morais, M. A. B., Chinaglia, M. et al. (2022). Gut microbiome of the largest living rodent harbors unprecedented enzymatic systems to degrade plant polysaccharides. Nat Commun. 13: 629. doi: 10.1038/s41467-022-28310-y.

Cani, P. D., Bibiloni, R., Knauf, C., Waget, A., Neyrinck, A. M., Delzenne, N. M. et al. (2008). Changes in gut microbiota control metabolic endotoxemia-induced inflammation in high-fat diet-induced obesity and diabetes in mice. Diabetes 57(6): 1470–1481.

Cetta, F. (1991). The role of bacteria in pigment gallstone disease. Ann Surg. 213: 315–26.

Chauhan, N. S., Nain, S. and Sharma, R. (2017). Identification of arsenic resistance genes from marine sediment metagenome. Indian J. Microbiol. 2017 Sep; 57(3): 299–306. doi: 10.1007/s12088-017-0658-0. Epub 2017 Jul 5. PMID: 28904414; PMCID: PMC5574776.

Chevalier, C., Stojanović, O., Colin, D. J., Suarez-Zamorano, N., Tarallo, V., Veyrat-Durebex, C. et al. (2015). Gut microbiota orchestrates energy homeostasis during cold. Cell. 163(6): 1360–1374.

Church, D. L., Cerutti, L., Gürtler, A., Griener, T., Zelazny, A. and Emler, S. (2020). Performance and application of 16S rRNA gene cycle sequencing for routine identification of bacteria in the clinical microbiology laboratory. Clin. Microbiol. Rev. 2020 Sep 9; 33(4): e00053–19. doi: 10.1128/CMR.00053-19. PMID: 32907806.

Cohen, H. Y., Miller, C., Bitterman, K. J., Wall, N. R., Hekking, B., Kessler, B. et al. (2004). Calorie restriction promotes mammalian cell survival by inducing the SIRT1 deacetylase. Science 305(5682): 390–392.

Collado, M. C., Derrien, M., Isolauri, E., de Vos, W. M. and Salminen, S. (2007). Intestinal integrity and Akkermansia muciniphila, a mucin-degrading member of the intestinal microbiota present in infants, adults, and the elderly. Applied and Environmental Microbiology 73(23): 7767–7770.

Collado, M. C., Isolauri, E., Laitinen, K. and Salminen, S. (2010). Effect of mother's weight on infant's microbiota acquisition, composition, and activity during early infancy: A prospective follow-up study initiated in early pregnancy. The American Journal of Clinical Nutrition 92(5): 1023–1030.

Cuscó, A., Pérez, D., Viñes, J., Fàbregas, N. and Francino, O. (2021). Long-read metagenomics retrieves complete single-contig bacterial genomes from canine feces. BMC Genomics. 2021 May 6; 22(1): 330. doi: 10.1186/s12864-021-07607-0. PMID: 33957869; PMCID: PMC8103633.

Cuscó, A., Pérez, D., Viñes, J., Fàbregas, N. and Francino O. (2022). Novel canine high-quality metagenome-assembled genomes, prophages and host-associated plasmids provided by long-read metagenomics together with Hi-C proximity ligation. Microb Genom. 2022 Mar; 8(3): 000802. doi: 10.1099/mgen.0.000802. PMID: 35298370; PMCID: PMC9176287.

den Besten, G., van Eunen, K., Groen, A. K., Venema, K., Reijngoud, D. J. and Bakker, B. M. (2013). The role of short-chain fatty acids in the interplay between diet, gut microbiota, and host energy metabolism. J. Lipid Res. 2013 Sep; 54(9): 2325–40. doi: 10.1194/jlr.R036012. Epub 2013 Jul 2. PMID: 23821742; PMCID: PMC3735932.

den Besten, G., Lange, K., Havinga, R., van Dijk, T. H., Gerding, A., van Eunen, K. et al. (2013). Gut-derived short-chain fatty acids are vividly assimilated into host carbohydrates and lipids. American Journal of Physiology-Gastrointestinal and Liver Physiology 305(12): G900–G910. https://doi.org/10.1152/ajpgi.00265.2013.

Deo, P. N. and Deshmukh, R. (2019). Oral microbiome: Unveiling the fundamentals. J. Oral Maxillofac Pathol. 2019 Jan-Apr; 23(1): 122–128. doi: 10.4103/jomfp.JOMFP_304_18. PMID: 31110428; PMCID: PMC6503789.

Di Tommaso, N., Gasbarrini, A. and Ponziani, F. R. (2021). Intestinal barrier in human health and disease. Int. J. Environ. Res. Public Health. 2021 Dec 6; 18(23): 12836. doi: 10.3390/ijerph182312836. PMID: 34886561; PMCID: PMC8657205.

Dyksma, S. and Gallert, C. (2022). Effect of magnetite addition on transcriptional profiles of syntrophic Bacteria and Archaea during anaerobic digestion of propionate in wastewater sludge. Environ Microbiol Rep. 2022 Aug; 14(4): 664–678. doi: 10.1111/1758-2229.13080. Epub 2022 May 25. PMID: 35615789.

Falk, M. W., Song, K.-G., Matiasek, M. G. and Wuertz, S. (2009). Microbial community dynamics in replicate membrane bioreactors—natural reproducible fluctuations. Water Res. 43: 842–852. doi: 10.1016/j.watres.2008.11.021.

Forster, S. C., Kumar, N., Anonye, B. O., Almeida, A., Viciani, E., Stares, M. D. et al. (2019). A human gut bacterial genome and culture collection for improved metagenomic analyses. Nat Biotechnol. 2019 Feb; 37(2): 186–192. doi: 10.1038/s41587-018-0009-7. Epub 2019 Feb 4. PMID: 30718869; PMCID: PMC6785715.

Fortea, M., Albert-Bayo, M., Abril-Gil, M., Ganda Mall, J. P., Serra-Ruiz, X., Henao-Paez, A. et al. (2021). Present and future therapeutic approaches to barrier dysfunction. Front Nutr. 2021 Oct 28; 8: 718093. doi: 10.3389/fnut.2021.718093. PMID: 34778332; PMCID: PMC8582318.

García Álvaro, A., Ruiz Palomar, C., Valenzuela, E. I., Hermosilla Redondo, D., Muñoz Torre, R. and de Godos Crespo, I. (2024). Microbial analysis of anaerobic digester reveals prevalence of manure microbiota. Journal of Water Process Engineering 60: 105162. https://doi.org/10.1016/j.jwpe.2024.105162.

Gerardi, M. H. (2003). Wastewater Microbiology Series. John Wiley & Sons Inc.; New Jersey, NJ, USA. The Microbiology of Anaerobic Digesters; pp. 121–125.

Ghatak, S., Muthukumaran, R. B. and Nachimuthu, S. K. (2013). A simple method of genomic DNA extraction from human samples for PCR-RFLP analysis. J. Biomol. Tech. 2013 Dec; 24(4): 224–31. doi: 10.7171/jbt.13-2404-001. PMID: 24294115; PMCID: PMC3792701.

Gillespie, D. E., Brady, S. F., Bettermann, A. D., Cianciotto, N. P., Liles, M. R., Rondon, M.R. et al. (2002). Isolation of antibiotics turbomycin a and B from a metagenomic library of soil microbial DNA. Appl Environ Microbiol. 68(9): 4301–6. doi: 10.1128/AEM.68.9.4301-4306.2002. PMID: 12200279; PMCID: PMC124076.

Gophna, U., Konikoff, T. and Nielsen, H. B. (2017). Oscillospira and related bacteria - From metagenomic species to metabolic features. Environ Microbiol. 2017 Mar; 19(3): 835–841. doi: 10.1111/1462-2920.13658. Epub 2017 Feb 3. PMID: 28028921.

Greer, R. L., Dong, X., Moraes, A. C., Zielke, R. A., Fernandes, G. R., Peremyslova, E. et al. (2016). Akkermansia muciniphila mediates negative effects of IFNγ on glucose metabolism. Nat Commun. 2016 Nov 14; 7: 13329. doi: 10.1038/ncomms13329. PMID: 27841267; PMCID: PMC5114536.

Grover, R., Burse, S., Shankrit, S., Aggarwal, A., Kirty, K., Narta, K. et al. (2019), Myg1 exonuclease couples the nuclear and mitochondrial translational programs through RNA processing. Nucleic Acids Research 47(11): 5852–5866, https://doi.org/10.1093/nar/gkz371.

Gulyaeva, A., Garmaeva, S., Ruigrok, R.A.A.A., Wang, D., Riksen, N. P., Netea, M. G. et al. (2022). Discovery, diversity, and functional associations of crAss-like phages in human gut metagenomes from four Dutch cohorts. Cell Rep. 2022 Jan 11; 38(2): 110204. doi: 10.1016/j.celrep.2021.110204. PMID: 35021085.

Handelsman, J., Rondon, M. R., Brady, S. F., Clardy, J. and Goodman, R. M. (1998). Molecular biological access to the chemistry of unknown soil microbes: a new frontier for natural products. Chem Biol. 5(10): R245–9. doi: 10.1016/s1074-5521(98)90108-9. PMID: 9818143.

Hao, L., Michaelsen, T. Y., Singleton, C. M., Dottorini, G., Kirkegaard, R. H., Albertsen, M. et al. (2020). Novel syntrophic bacteria in full-scale anaerobic digesters revealed by genome-centric metatranscriptomics. ISME J. 14: 906–918. doi: 10.1038/s41396-019-0571-0.

Johnson, N. W., Jayasekara, P. and Amarasinghe, A. A. H. K. (2011). Squamous cell carcinoma and precursor lesions of the oral cavity: epidemiology and aetiology. Periodontology 57(1): 19–37. https://doi.org/10.1111/j.1600-0757.2011.00401.x.

Johnson, D. E., Burtness, B., Leemans, C. R., Lui, V. W. Y., Bauman, J. E. and Grandis, J. R. (2020). Head and neck squamous cell carcinoma. Nat Rev Dis Primers. 2020 Nov 26; 6(1): 92. doi: 10.1038/s41572-020-00224-3.

Johnson, J. S., Spakowicz, D. J., Hong, B. Y., Petersen, L. M., Demkowicz, P., Chen, L. et al. (2019). Evaluation of 16S rRNA gene sequencing for species and strain-level microbiome analysis. Nat Commun. 2019 Nov 6; 10(1): 5029. doi: 10.1038/s41467-019-13036-1. PMID: 31695033; PMCID: PMC6834636.

Johnston, K. L., Swift, N. C., Forster-van Hijfte, M., Rutgers, H. C., Lamport, A., Ballèvre, O. et al. (2001). Comparison of the bacterial Flora of the duodenum in healthy cats and cats with signs of gastrointestinal tract disease. J. Am. Vet. Med. Assoc. 218: 48–51. doi: 10.2460/javma.2001.218.48.

Kirkegaard, R. H., McIlroy, S. J., Kristensen, J. M., Nierychlo, M., Karst, S. M., Dueholm, M. S. et al. (2017). The impact of immigration on microbial community composition in full-scale anaerobic digesters. Sci. Rep. 7: 9343. doi: 10.1038/s41598-017-09303-0.

Kong, D., Zhang, K., Liang, J., Gao, W. and Du, L. (2019). Methanogenic community during the anaerobic digestion of different substrates and organic loading rates. Microbiologyopen. 2019 May; 8(5): e00709. doi: 10.1002/mbo3.709. Epub 2018 Aug 15. PMID: 30112808; PMCID: PMC6528610.

Kose, S. H., Grice, K., Orsi, W. D., Ballal, M., Coolen, M. J. L. (2018). Metagenomics of pigmented and cholesterol gallstones: the putative role of bacteria. Sci. Rep. 2018 Jul 25; 8(1): 11218. doi: 10.1038/s41598-018-29571-8. Erratum in: Sci Rep. 2020 Mar 4; 10(1): 4347. PMID: 30046045; PMCID: PMC6060111.

Kostic, A. D., Chun, E., Robertson, L., Glickman, J. N., Gallini, C. A., Michaud, M. et al. (2013). *Fusobacterium nucleatum* potentiates intestinal tumorigenesis and modulates the tumor-immune microenvironment. Cell Host Microbe. 2013 Aug 14; 14(2): 207–15. doi: 10.1016/j.chom.2013.07.007. PMID: 23954159; PMCID: PMC3772512.

Li, C., Hao, L., Lü, F., Duan, H., Zhang, H. and He, P. (2022). Syntrophic acetate-oxidizing microbial consortia enriched from full-scale mesophilic food waste anaerobic digesters showing high biodiversity and functional redundancy. mSystems. 2022 Oct 26; 7(5): e0033922. doi: 10.1128/msystems.00339-22. Epub 2022 Sep 8. PMID: 36073802; PMCID: PMC9600251.

Li, J., Xie, S., Ahmed, S., Wang, F., Gu, Y., Zhang, C. et al. (2017). Antimicrobial activity and resistance: influencing factors. Front Pharmacol. 2017 Jun 13; 8: 364. doi: 10.3389/fphar.2017.00364. PMID: 28659799; PMCID: PMC5468421.

Li, Y., Ma, J., Yuan, H. and Li, X. (2022). Effects of feeding regimes on process performance and microbial community structure in anaerobic semi-continuously stirred tank reactors treating corn stover. Waste Biomass Valorization. 13: 1003–1014. doi: 10.1007/s12649-021-01573-0.

Lim, J. W., Park, T., Tong, Y. W. and Yu, Z. (2020). The microbiome driving anaerobic digestion and microbial analysis. Advances in Bioenergy 5: 1–61. doi: 10.1016/bs.aibe.2020.04.001. Epub 2020 May 4. PMCID: PMC7198183.

López, J. M. S. and Bogaert, I. N. A. V. (2021). Microbial fatty acid transport proteins and their biotechnological potential Biotechnol. Bioeng. 118(6): 2184–2201.

Madhavi, A., Hingane, S., Srivastav, R., Joshi, N., Subramani, C., Muthumohan, R. et al. (2017). A Screen for novel hepatitis C virus RdRp inhibitor identifies a broad-spectrum anti-viral compound. Scientific Reports 7(1): 5816. https://doi.org/10.1038/s41598-017-04449-3.

Maki, T. (1996). Pathogenesis of calcium bilirubinate gallstone: Role of *E. coli*, beta glucuronidase and coagulation by inorganic ions, polyelectrolytes, and agitation. Ann Surg. 164: 90–100.

Marbouty, M., Thierry, A., Millot, G. A. and Koszul, R. (2021). MetaHiC phage-bacteria infection network reveals active cycling phages of the healthy human gut. Elife. 2021 Feb 26; 10: e60608. doi: 10.7554/eLife.60608. PMID: 33634788; PMCID: PMC7963479.

Mauky, E., Jacobi, H. F., Liebetrau, J. and Nelles, M. (2015). Flexible biogas production for demand-driven energy supply—feeding strategies and types of substrates. Bioresour. Technol. 178: 262–269. doi: 10.1016/j.biortech.2014.08.123.

McCallum, G. E., Rossiter, A. E., Quraishi, M. N., Iqbal, T. H., Kuehne, S. A. and van Schaik, W. (2023). Noise reduction strategies in metagenomic chromosome confirmation capture to link antibiotic resistance genes to microbial hosts. Microb Genom. 2023 Jun; 9(6): mgen001030. doi: 10.1099/mgen.0.001030. PMID: 37272920; PMCID: PMC10327510.

McIlvanna, E., Linden, G. J., Craig, S. G., Lundy, F. T. and James, J. A. (2021). *Fusobacterium nucleatum* and oral cancer: a critical review. BMC Cancer. 2021 Nov 13; 21(1): 1212. doi: 10.1186/s12885-021-08903-4. PMID: 34774023; PMCID: PMC8590362.

McInerney, M. J., Struchtemeyer, C..G., Sieber, J., Mouttaki, H., Stams, A. J. M., Schink, B. et al. (2008). Physiology, ecology, phylogeny, and genomics of microorganisms capable of syntrophic metabolism. Ann. NY Acad. Sci. 1125: 58–72. doi: 10.1196/annals.1419.005.

Meng, X., Cao, Q., Sun, Y., Huang, S., Liu, X. and Li, D. (2022). 16S rRNA genes- and metagenome-based confirmation of syntrophic butyrate-oxidizing methanogenesis enriched in high butyrate loading. Bioresour Technol. 2022 Feb; 345: 126483. doi: 10.1016/j.biortech.2021.126483. Epub 2021 Dec 2. PMID: 34864185.

Milani, C., Alessandri, G., Mancabelli, L., Mangifesta, M., Lugli, G. A., Viappiani, A. et al. (2020). Multi-omics approaches to decipher the impact of diet and host physiology on the mammalian gut microbiome. Appl. Environ. Microbiol. 86. doi: 10.1128/AEM.01864-20.

Miller, S., Naccache, S. N., Samayoa, E., Messacar, K., Arevalo, S., Federman, S. et al. (2019). Laboratory validation of a clinical metagenomic sequencing assay for pathogen detection in cerebrospinal fluid. Genome Res. 2019 May; 29(5): 831–842. doi: 10.1101/gr.238170.118. PMID: 30992304; PMCID: PMC6499319.

Minarovits, J. (2021). Anaerobic bacterial communities associated with oral carcinoma: Intratumoral, surface-biofilm and salivary microbiota. Anaerobe 2021 Apr; 68: 102300. doi: 10.1016/j.anaerobe.2020.102300. Epub 2020 Nov 24. PMID: 33246097.

Mulat, D. G., Jacobi, H. F., Feilberg, A., Adamsen, A. P. S., Richnow, H.-H. and Nikolausz, M. (2016). Changing feeding regimes to demonstrate flexible biogas production: effects on process performance, microbial community structure, and methanogenesis pathways. Appl. Env. Microb. 82: 438–449. doi: 10.1128/AEM.02320-15.

Nair, R. R., Rangaswamy, B., Sarojini, B. S. I. and Joseph, V. (2020). Anaerobic ammonia-oxidizing bacteria in tropical bioaugmented zero water exchange aquaculture ponds. Environ Sci. Pollut. Res. Int. 2020 Apr; 27(10): 10541–10552. doi: 10.1007/s11356-020-07663-1. Epub 2020 Jan 15. PMID: 31940146.

Nnadozie, C. F. and Odume, O. N. (2019). Freshwater environments as reservoirs of antibiotic resistant bacteria and their role in the dissemination of antibiotic resistance genes. Environ Pollut. 2019 Nov; 254(Pt B): 113067. doi: 10.1016/j.envpol.2019.113067. Epub 2019 Aug 21. PMID: 31465907.

Op den Camp, H. J. M., Kartal, B., Guven, D., van Niftrik, L. A. M. P., Haaijer, S. C. M., van der Star, W. R. L. et al. (2006). Global impact and application of the anaerobic ammonium-oxidizing (anammox) bacteria. Biochemical Society Transactions 34(Pt 1): 174–178. https://doi.org/10.1042/BST0340174.

Oshiki, M., Takaki, Y., Hirai, M., Nunoura, T., Kamigaito, A. and Okabe, S. (2022). Metagenomic analysis of five phylogenetically distant anammox bacterial enrichment cultures. Microbes Environ. 37(3): ME22017. doi:10.1264/jsme2.ME22017.

Patangia, D. V., Anthony Ryan, C., Dempsey, E., Paul Ross, R. and Stanton, C. (2022). Impact of antibiotics on the human microbiome and consequences for host health. Microbiologyopen 2022 Feb; 11(1): e1260. doi: 10.1002/mbo3.1260. PMID: 35212478; PMCID: PMC8756738.

Pereira, R., Oliveira, J. and Sousa, M. (2020). Bioinformatics and computational tools for next-generation sequencing analysis in clinical genetics. J. Clin. Med. 2020 Jan 3; 9(1): 132. doi: 10.3390/jcm9010132. PMID: 31947757; PMCID: PMC7019349.

Pilarska, A. A., Kulupa, T., Kubiak, A., Wolna-Maruwka, A., Pilarski, K. and Niewiadomska, A. (2023). Anaerobic digestion of food waste—a short review. Energies 16(15): 5742. https://doi.org/10.3390/en16155742.

Radhakrishanan, A., Kakkar, P., Tandon, C. and Srivastav, R. (2022). Fecal microbiome transplantation: an offhand recipe for microbiome therapeutics. pp. 246–256. *In*: Glibetic, M. (ed.), Comprehensive Gut Microbiota, vol. 3. Elsevier. https://dx.doi.org/10.1016/B978-0-12-819265-8.00001-2. 13/1/2022.

Radhakrishnan, A., Sharma, C., Malviya, V. N. and Srivastav, R. (2024). Deciphering the molecular functionality of Cdc45 in replisomal complex. Biochemistry and Biophysics Reports 37: 101643, ISSN 2405-5808, https://doi.org/10.1016/j.bbrep.2024.101643.

Radhakrishnan, A., Kapil, T., Kapardar, R. and Srivastav, R. (2023). Microbiome additive therapy for the human health, Editor(s): Nar Singh Chauhan, Suneel Kumar, Microbiome Therapeutics, Academic Press, Pages 41–61, https://doi.org/10.1016/B978-0-323-99336-4.00011-2.

Ramos, G. P. and Papadakis, K. A. (2019). Mechanisms of disease: inflammatory bowel diseases. Mayo Clin Proc. 2019 Jan; 94(1): 155–165. doi: 10.1016/j.mayocp.2018.09.013. PMID: 30611442; PMCID: PMC6386158.

Razzaq, A., Shamsi, S., Ali, A., Ali, Q., Sajjad, M., Malik, A. et al. (2019). Microbial proteases applications. Front Bioeng. Biotechnol. 2019 Jun 12; 7: 110. doi: 10.3389/fbioe.2019.00110. PMID: 31263696; PMCID: PMC6584820.

Rodrigues, V. F., Elias-Oliveira, J., Pereira, Í. S., Pereira, J. A., Barbosa, S. C., Machado, M. S. G. et al. (2022). *Akkermansia muciniphila* and gut immune system: a good friendship that attenuates inflammatory bowel disease, obesity, and diabetes. Frontiers in Immunology 13. https://doi.org/10.3389/fimmu.2022.934695.

Rojas, C. A., Gardy, J., Eisen, J. A. and Ganz, H. H. (2023). Recovery of 52 bacterial genomes from the fecal microbiome of the domestic cat (Felis catus) using Hi-C proximity ligation and shotgun metagenomics. Microbiol Resour Announc. 2023 Oct 19; 12(10): e0060123. doi: 10.1128/MRA.00601-23. Epub 2023 Sep 11. PMID: 37695121; PMCID: PMC10586161.

S. P. Galhano, B., G. Ferrari, R., Panzenhagen, P., S. de Jesus, A. C. and A. Conte-Junior, C. (2021). Antimicrobial resistance gene detection methods for bacteria in animal-based foods: a brief review of highlights and advantages. Microorganisms 2021 Apr 26; 9(5): 923. doi: 10.3390/microorganisms9050923. PMID: 33925810; PMCID: PMC8146338.

Sakai, S. A., Aoshima, M., Sawada, K., Horasawa, S., Yoshikawa, A., Fujisawa, T. et al. (2022). Fecal microbiota in patients with a stoma decreases anaerobic bacteria and alters taxonomic and functional diversities. Front Cell Infect Microbiol. 2022 Sep 14; 12: 925444. doi: 10.3389/fcimb.2022.925444. PMID: 36189350; PMCID: PMC9515963.

Shih, C. T., Yeh, Y. T., Lin, C. C., Yang, L. Y. and Chiang, C. P. (2020). *Akkermansia muciniphila* is negatively correlated with hemoglobin A1c in refractory diabetes. Microorganisms 2020 Sep 5; 8(9): 1360. doi: 10.3390/microorganisms8091360. PMID: 32899513; PMCID: PMC7565276.

Somerville, V., Lutz, S., Schmid, M., Frei, D., Moser, A., Irmler, S. et al. (2019). Long-read based de novo assembly of low-complexity metagenome samples results in finished genomes and reveals insights into strain diversity and an active phage system. BMC Microbiol. 2019 Jun 25; 19(1): 143. doi: 10.1186/s12866-019-1500-0. PMID: 31238873; PMCID: PMC6593500.

Sousa, D. Z., Smidt, H., Alves, M. M. and Stams, A. J. M. (2009). Ecophysiology of syntrophic communities that degrade saturated and unsaturated long-chain fatty acids: ecophysiology of syntrophic LCFA degradation. FEMS Microbiol. Ecol. 68: 257–272. doi: 10.1111/j.1574-6941.2009.00680.x.

Srivastav, R., Kumar, D., Grover, A., Singh, A., Manjasetty, B., Sharma, R. et al. (2014). unique subunit packing in mycobacterial nanoRNase leads to alternate substrate recognitions in DHH phosphodiesterases. Nucleic Acids Research 42(12): 7894–7910. https://doi.org/10.1093/nar/gku425.

Srivastav, R., Sharma, R., Tandon, S. and Tandon, C. (2019). Role of DHH superfamily proteins in nucleic acids metabolism and stress tolerance in prokaryotes and eukaryotes. Int. J. Biol. Macromol. 127: 66–75. PMID: 30578903. https://doi.org/10.1016/j.ijbiomac.2018.12.123.

Srivastav, R., Singh, A., Jangir, P. K., Kumari, C., Muduli, S. and Sharma, R. (2013). Genome sequence of Staphylococcus massiliensis strain S46 isolated from surface of healthy human skin. Genome Announc 1(4): e00553-13 PMID: 23929469 https://doi.org/10.1128/genomeA.00553-13.

Srivastav, R. and Suneja, G. (2019). Recent advances in microbial genome sequencing. *In*: Tripathi, V., Kumar, P., Tripathi, P., Kishore, A. and Kamle, M. (eds.). Microbial Genomics in Sustainable Agroecosystems. Springer Ed.1 Vol 2 Chapt 8 ISBN 9789813298606 publisher https://doi.org/10.1007/978-981-32-9860-6.

Stewart, L., Oesterle, A. L., Erdan, I., Griffiss, J. M. and Way, L. W. (2002). The pathogenesis of pigment gallstones in western societies: the central role of bacteria. J. Gastrointest Surg. 6: 891–904.

Stewart, L., Smith, A. L., Pellegrini, C. A., Motson, R. W. and Way, L. W. (1987). Pigment gallstones form as a composite of bacterial micro-colonies and pigment solids. Ann Surg. 206: 242–250.

Suarez, C,, Dalcin Martins, P., Jetten, M. S. M., Karačić, S., Wilén, B. M., Modin, O. et al. (2022). Metagenomic evidence of a novel family of anammox bacteria in a subsea environment. Environ Microbiol. 2022 May; 24(5): 2348–2360. doi: 10.1111/1462-2920.16006. Epub 2022 Apr 18. PMID: 35415863; PMCID: PMC9325076.

Suneja, G. and Srivastav, R. (2021). Impact of microbial genome sequencing advancements in understanding extremophiles. Book -Extreme Environments, Unique Ecosystems – Amazing Microbes, Ed.- Anita Pandey, & Avinash Sharma, CRC Press. ISBN 9780367350161 Ed. 1,pp 13.

Tao, Y., Xun, F., Zhao, C., Mao, Z., Li, B., Xing, P. et al. (2023). Improved assembly of metagenome-assembled genomes and viruses in Tibetan Saline Lake sediment by HiFi metagenomic sequencing. Microbiol Spectr. 2023 Feb 14; 11(1): e0332822. doi: 10.1128/spectrum.03328-22. Epub 2022 Dec 8. PMID: 36475839; PMCID: PMC9927493.

Van Berkum, N. L., Lieberman-Aiden, E., Williams, L., Imakaev, M., Gnirke, A., Mirny, L. A. et al. (2010). Hi-C: a method to study the three-dimensional architecture of genomes. J. Vis. Exp. 39: 1869. doi: 10.3791/1869.

Vanwonterghem, I., Jensen, P. D., Dennis, P. G., Hugenholtz, P., Rabaey, K. and Tyson, G. W. (2014). Deterministic processes guide long-term synchronised population dynamics in replicate anaerobic digesters. ISME J. 8: 2015–2028. doi: 10.1038/ismej.2014.50.

Vrieze, J. D., Hennebel, T., Boon, N. and Verstraete, W. (2012). Methanosarcina: The rediscovered methanogen for heavy duty biomethanation. Bioresour. Technol. 112: 1–9. doi: 10.1016/j.biortech.2012.02.079.

Vrieze, J. D., Verstraete, W. and Boon, N. (2013). Repeated pulse feeding induces functional stability in anaerobic digestion. Microb. Biotechnol. 6: 414–424. doi: 10.1111/1751-7915.12025.

Willeghems, G. and Buysse, J. (2016). Changing old habits: the case of feeding patterns in anaerobic digesters. Renew. Energ. 92: 212–221. doi: 10.1016/j.renene.2016.01.081.

Xie, F., Jin, W., Si, H., Yuan, Y., Tao, Y., Liu, J. et al. (2021). An integrated gene catalog and over 10,000 metagenome-assembled genomes from the gastrointestinal microbiome of ruminants. Microbiome 9: 137. doi: 10.1186/s40168-021-01078-x.

Xu, R., Kolton, M., Tao, W., Sun, X., Su, P., Huang, D. et al. (2023). Anaerobic selenite-reducing bacteria, and their metabolic potentials in Se-rich sediment revealed by the combination of DNA-stable isotope probing, metagenomic binning, and metatranscriptomics. J. Hazard Mater. 2023 Sep 5; 457: 131834. doi: 10.1016/j.jhazmat.2023.131834. Epub 2023 Jun 11. PMID: 37327607.

Xue Deng, Chengyuan Su, Xiaoen Luo, Xiaoyue Liu, Zhu Liang, Fengxiu Ding et al. (2022). Performance and metagenomic analysis of a micro-aerated anaerobic reactor for biological sulfate-rich wastewater treatment. Journal of Environmental Chemical Engineering 10(3): 107769, ISSN 2213-3437.

Young, W., Moon, C. D., Thomas, D. G., Cave, N. J. and Bermingham, E. N. (2016). Pre- and post-weaning diet alters the faecal metagenome in the cat with differences in vitamin and carbohydrate metabolism gene abundances. Sci Rep. 6: 34668. doi: 10.1038/srep34668.

Yulandi, A., Suwanto, A., Waturangi, D. E. and Wahyudi, A. T. (2020). Shotgun metagenomic analysis reveals new insights into bacterial community profiles in tempeh. BMC Res Notes. 2020 Dec 11; 13(1): 562. doi: 10.1186/s13104-020-05406-6. PMID: 33308279; PMCID: PMC7731626.

Zhao, R., Le Moine Bauer, S. and Babbin, A. R. (2023). "Candidatus Subterrananammoxibiaceae," a new anammox bacterial family in globally distributed marine and terrestrial subsurfaces. Appl. Environ. Microbiol. 89(8): e0080023. doi:10.1128/aem.00800-23.

Zheng, M., Han, R., Yuan, Y., Xing, Y., Zhang, W., Sun, Z. et al. (2023). The role of *Akkermansia muciniphila* in inflammatory bowel disease: Current knowledge and perspectives. Front Immunol. 2023 Jan 6; 13: 1089600. doi: 10.3389/fimmu.2022.1089600. PMID: 36685588; PMCID: PMC9853388.

Zhou, K. (2017). Strategies to promote abundance of *Akkermansia muciniphila*, an emerging probiotics in the gut, evidence from dietary intervention studies. J. Funct Foods. 2017 Jun; 33: 194–201. doi: 10.1016/j.jff.2017.03.045. Epub 2017 Mar 29. PMID: 30416539; PMCID: PMC6223323.

Ziels, R. M., Beck, D. A. C. and Stensel, H. D. (2017). Long-chain fatty acid feeding frequency in anaerobic codigestion impacts syntrophic community structure and biokinetics. Water Res. 117: 218–229. doi: 10.1016/j.watres.2017.03.060.

Chapter 9

Exploring the Microbiome and Anaerobes in Soil for Sustainable Agriculture

Nanditha Pramod

9.1 Introduction

Land, ground, also called Earth, comprises the terrestrial part of the planet Earth, that is not under water. The face of the land is covered by an enormous grainy layer called soil, which is an essential component of land. Land houses human beings, animals, and other land species, and soil is the home for diverse microorganisms. Apart from the numerous life forms the soil consists of a complex mixture of minerals, organic matter, water, and air.

Microorganisms are cosmopolitan in existence. The air we breathe, the water we drink, the food we eat, the land we walk on, and even our bodies have a plethora of microorganisms residing and multiplying every minute. Among all these, the soil is known to be the largest reservoir of microorganisms on Earth. Survival in the soil poses several challenges despite of which a gram of soil will contain organisms from all three domains of life, along with viruses (Prabhakaran et al. 2022). The majority of the Earth's biomass is made up of microorganisms living in the soil and deep surfaces, following plants. This microbial community residing in a comprehensible habitat that has its physio-chemical properties is collectively called the microbiome (Banerjee and var der Heijden 2023). The overwhelming majority of the living biomass in the soil is primarily of bacteria accounting for about 15%. Still, there are also more fungi and archaea- 2 and 1%, respectively, than animals which is only 0.3%. In virtually every soil, fungi and bacteria predominate over protists and

School of Medical Sciences, University of Hyderabad, Hyderabad, Telangana - 500046.
Email: nandithapramod07@gmail.com

DOI: 10.1201/9781003506898-9

archaea in terms of biomass, though viruses also reside in the soil, their proportion is negligible (Bar-On et al. 2018).

Among the bacterial species inhabiting the soil, some of them are anaerobic bacteria. They are extensively present in the anoxic conditions, mostly seen in aquatic soils, flooded rice paddy fields, subzero temperatures in glaciers, permafrost, polar regions (Honma et al. 2023) and even extreme temperatures as in deep sea vents, volcanoes, and hot springs (Lu and Imlay 2021). Humanly inaccessible and extreme environments are the home for several anaerobes. Adding to this, is the difficulty in culturing anaerobes, leaving several aspects on the functions of anaerobic bacteria in the ecosystem unexplored (Patidar and Prakash 2022). Studies have shown that, excess oxygen curbs their growth, while they can tolerate minor bursts of oxygen through well-developed enzymatic mechanisms and later resume their growth (Lu and Imlay 2021). Various industrial applications, namely, fermentation, biofuel production, synthesis of vaccines, vitamins, and dairy employ anaerobic bacteria. Currently, anaerobes are also being used in agriculture in processes like biological soil disinfestation aiding in sustainable agricultural practices (Patidar and Prakash 2022).

In a nutshell, all these soil-dwellers which make up the microbiome are critical for soil health. Only healthy and nutrient-rich soil supports plant productivity, thus improving the quality of life of all organisms on the planet, including human beings. With the increase in human population, the demands for global food production have also sky-rocketed. Soil being the foundation for agriculture determines the outputs depending on the quality of the soil. This in turn can be obtained by understanding the soil microbiome and incorporating this understanding into sustainable agricultural practices and other applications (Tosi et al. 2020). The soil microbiome influences plant growth and productivity by determining the fertility of the soil, aiding in stress tolerance, and providing the required nutrients, oxygen, and water (Hartmann and Six 2023). The important roles played by microbiomes including anaerobes in ensuring soil health make it a compelling need to understand and explore the soil microbiome in greater depth.

9.2 Unearthing the Soil Microbiome

9.2.1 Soil Types and the Associated Microbiome

Soil microbiome is composed of a diverse population of bacteria, fungi, archaea, viruses, and protozoa. The diversity of the microbiome composition varies across the globe. For example, soil bacterial diversity is the most at the mid-latitudes and minimal at the equator and the poles (Bahram et al. 2018). The microbial diversity in soil differs from one soil type to another. The main factor influencing this trend is soil moisture. Soils with the optimum moisture content house an enormous community of microorganisms and determine the biochemical soil activity (Borowik and Wyszkowska 2016). Similarly, the anoxic and reduced environments as seen in flooded paddy fields, house several anaerobic bacteria and archaea, notably methanogens (Honma et al. 2023).

Billions of hectares of land on Earth are occupied by forests. They are the major carbon sinks and nutrient sources on the planet and are the most productive among terrestrial ecosystems (Onyango et al. 2023). Studies on the fungal diversity in forest soils are numerous, as they are responsible for the carbon cycle (Li et al. 2022). Basidiomycota was the most prevalent fungal community in forests. A constant shift in abundances of Ascomycota and Zygomycota was also reported in forest soils (Santalahti et al. 2016). Though many reports are focused on fungal communities of forest soils, it should be noted that bacteria are also major players in maintaining the forest ecosystem (Lladó et al. 2017). It was reported that Proteobacteria, Firmicutes, Acidobacteria, Bacteroidetes, and Actinobacteria are the major bacterial phyla inhabiting forest soils (Lauber et al. 2009). While Proteobacteria predominated among the bacteria, Crenarchaeota was the most abundant archaeal phyla in forest soils (Onyango et al. 2023).

Grasslands are those areas of land that have grass species predominantly present when compared to other plants. A significant amount of the world's agricultural terrain is made up of grasslands (Do et al. 2023). Studies have shown that numerous bacteria, fungi, and other microbes are the major inhabitants of grasslands. The bacterial phyla Proteobacteria and Actinobacteria were abundant in grassland soils. However, the genus *Bradyrhizhobium* was predominant in soil (Do et al. 2023). It was also reported that arbuscular mycorrhizal fungi associated with plants are found in grasslands (Faghihinia et al. 2020). Also common in grasslands are the single-celled organisms protists (Fiore-Donno et al. 2019).

Drylands called deserts cover about 35% of the planet, and currently, there is an increase in the aridity cover caused majorly due to climate change. Microorganisms with the ability to tolerate harsh and arid conditions are the dominating communities in such lands. It was found that Firmicutes (bacteria), Ciliophora (protistan), and Sordariomycetes (fungi) and abundantly found with increasing aridity (Dong et al. 2024).

When it comes to tundra soil, microbial community and diversity are less prominent in the Arctic, Antarctic, and other cold areas with permafrost. A recent study reported that the layers of permafrost harbour Actinobacteria, Proteobacteria, Alphaproteobacteria, Acidobacteria, Bacteroidetes, and Verrucomicrobia. It was also revealed that methanogenic Euryarchaeota, Methanobacterium, and Methanosaeta were also found in tundra soils under study (Taş et al. 2018).

9.2.2 Trends in Soil Microbiome with the Soil Depth

Soil microbiome varies with the increasing depth of the soil. The characteristics of the soil and the site under study, namely pH, temperature, water content, and availability of organic carbon are key players in determining the composition of soil microbiome (Prabhakaran et al. 2022). Also, the water-filled pores in the soil with resources serve as hotspots for the survival of the microbiome (Prabhakaran et al. 2022). It was reported in a study that the microbial biomass decreased in abundance with the soil depth in forests and grasslands, and no noticeable change was seen in agricultural lands (van Leeuwen et al. 2017).

The diversity, abundance, and composition of the microbiome vary throughout the depth of the soil. It is quite established that the microbial richness significantly decreases with the increase in soil depth. This decrease was highly correlated to the decreasing organic matter as depth increased (Hao et al. 2021). Bacterial communities were found to be dominating in all depths (Hao et al. 2021). While Actinomycetes and Gram-positive bacteria were abundant in deeper soils, Gram-negative bacteria declined with increasing depth (van Leeuwen et al. 2017). The intermediate layers were also inhabited by Verrucomicrobia, Actinobacteria, Armatimonadetes, Chloroflexi, and Nitrospirae. Unlike other bacterial phyla like Acidobacteria, Gemmatimonadetes, Planctomycetes, and Fibrobacteres, Proteobacteria increased in abundance with increasing soil depth (Hao et al. 2021). Several genera of bacteria were found to decrease with the increasing depth, except for *Escherichia, Phyllobacterium*, and *Hyphomicrobium*. *Escherichia* belonging to phylum Proteobacteria dominated the deeper layers (Hao et al. 2021). The photoautotrophic Cyanobacteria also decreased in abundance in deeper soils (Tripathi et al. 2018).

Archaea are specialized in tolerating the anaerobic and low-nutrient conditions that are characteristic features of deep soil layers. The archaeal phylum Crenarchaeota was found in the intermediate soil layers, they were the least in the upper and lowermost layers (Hao et al. 2021). Among the common archaeal phyla like Euryarchaeota, Crenarchaeota, and Thaumarchaeota, Euryarchaeota is predominant in deeper soil layers (Feng et al. 2019).

The surface soil layers are mostly inhabited by abundant numbers of fungal communities. The diversity of fungi also follows the same trend as bacteria and decreases with the depth of the soil (Naylor et al. 2022). Soil houses the fungal phyla like Ascomycota, Basidiomycota, and Zygomycota. Ascomycota and Basidiomycota are known to decline with depth, while Zygomycota tends to increase (Li et al. 2020) Table 9.1.

Table 9.1. Microorganisms composing the soil microbiome in different layers of the soil.

Soil layers	Dominant microbial phyla in the soil microbiome		References
Surface and uppermost soil layer	Bacteria	Proteobacteria, Actinobacteria, Acidobacteria, Chloroflexi, Firmicutes, Planctomycetes, Gemmatimonadetes, Nitrospirae, Rokubacteria, Verrucomicrobia, Cyanobacteria	Jiang et al. 2023, Hao et al. 2021, Naylor et al. 2022
	Fungi	Ascomycota, Basidiomycota, Mortierellomycota, Chytridiomycota, Neocallimastigomycota	Jiang et al. 2023, Li et al. 2020
	Archaea	Crenarchaeota	Hao et al. 2021, Feng et al. 2019
Intermediate soil layer	Bacteria	Proteobacteria, Actinobacteria, Acidobacteria Firmicutes, Verrucomicrobia, Armatimonadetes, Chloroflexi, Nitrospirae, Planctomycetes	Hao et al. 2021, Naylor et al. 2022
	Fungi	Ascomycota, Zygomycota, Glomeromycota	Li et al. 2020
	Archaea	Crenarchaeota, Thaumarchaeota	Hao et al. 2021
Lowermost soil layer	Bacteria	Proteobacteria	Hao et al. 2021
	Fungi	Zygomycota	Li et al. 2020
	Archaea	Euryarchaeota, Thaumarchaeota	Feng et al. 2019

9.3 Factors Influencing the Soil Microbiome

The microbial composition in the soil is dependent on the physio-chemistry of the soil. These include soil properties like porosity, pH, aeration, moisture, salinity, and nutrient content. The physical and chemical properties of the soil are in turn influenced by the surrounding environmental factors including climate change, temperature, rainfall, agriculture, and land use (Abdul Rahman et al. 2021). These abiotic, biotic, and edaphic factors influence the growth and development of the microbial community in the soil. One of the major parameters is the pH of the soil (Prabhakaran et al. 2022). pH shapes the microbial community inhabiting any soil (Hao et al. 2021). The effect of pH on the soil microbiome composition is indirect as it alters the microbial load by affecting the nitrogen content in the soil (Wang et al. 2023). The availability of oxygen in the soil also plays an important role, as does the organic matter content, water saturation, and bulk density (Hao et al. 2021). The minerals in the soil, namely, phosphorus, potassium, magnesium, aluminium, iron, and calcium are major players in determining the soil microbiome. Along with this are the nitrate levels in the soil (Naylor et al. 2022). One study reported that the crop types and locations also affect the soil microbiome. The soil depth is a fundamental factor influencing the same. It was observed that the microbial abundance in the surface layers between the crops showed a significant change when compared to the lower depths (Hao et al. 2021). In the present day, climate change tends to cause the most devastating impact on soil microbiome. The changing climatic conditions around the globe have led to a significant rise in temperature and carbon dioxide levels, an increase in the episodes of drought and floods, soil warming, etc., which adversely affects the soil microbial composition (Jansson and Hofmockel 2020).

9.4 Interaction of the Soil Microbiome with Plants

Soil apart from being the home for all organisms is in fact the source of life for plants. The plant roots interact with the soil for water, nutrients, anchorage, and many more. The roots of the plant facilitate the growth and survival of numerous microorganisms, which include rhizobacteria, mycorrhiza, actinomycetes, blue-green algae, protozoans, etc. The well-defined zone which is the site for plant soil-microbiome interaction is referred to as the ‘rhizosphere’. The interactions between the plants and soil microbiome can be symbiotic, commensalism, or neutral (Chauhan et al. 2023). Comprising a convoluted network of plant roots, soil, and a diverse microbiome, the rhizosphere is known to be the most complex microhabitat (Hakim et al. 2021). The plant roots secrete compounds that act as a connecting network between the plant and the beneficial microbes (Zehra et al. 2021). The soil surrounding the plant root is provided with an ideal environment rich with nutrients and attractants so that it can nurture the microbiome. The compounds used for this purpose are sugars, amino acids, organic acids, flavonoids, phenolic compounds, etc. (Compant et al. 2019). Plant health and growth are challenged by several phytopathogens. With the understanding of the toxic effects of pesticides, the world has now moved towards the employment of microbial species of the rhizosphere for pest control. This is indeed a major step in sustainable agriculture (Zehra et al. 2021).

9.4.1 Beneficial Functions of the Plant-Soil Microbiome

The residents of the rhizosphere composed of the diversity of microorganisms help in maintaining the soil and plant health. The plant-associated microbiome is known to promote plant growth and development and also protects the plants from pest or parasitic infestation and soil-borne diseases (Chauhan et al. 2023). The plant-soil microbiome is involved in nutrient cycling pathways and also protects the plant from several stresses. It produces plant growth hormones (auxin, cytokinin, gibberellin), secondary metabolites, antibiotics, and compounds for alleviating stress (Hakim et al. 2021). Microorganisms belonging to the rhizosphere carry out the mechanism of pathogen control by the production of antimicrobials, and enzymes, and by improving plant resistance to diseases (Zehra et al. 2021). They are also known to induce a systemic response in plants through jasmonic acid and ethylene signalling pathways (Mou et al. 2020).

9.4.2 Harmful Side of Plant-Soil Microbiome Interaction

The rhizosphere also harbours several microorganisms, some of which can adversely affect plant growth and can often be life-threatening to the plants. They exhibit this harmful side by competing for nutrients and space which can also result in severe diseases. The chemical signals released by the plants as an attractant for microbes itself is utilized by several pathogens to reach the plant. Biotrophs and necrotrophs are the two groups of phytopathogens. While biotrophs let the plants survive even under infection, the latter kill the plant cells and take up nutrients. The major phytopathogens that deteriorate plant health are fungi and nematodes (Chauhan et al. 2023). Some bacterial species produce phytotoxic compounds which adversely affect the plant. To name a few, *Pseudomonas syringae*, *Xanthomonas*, *Ralstonia solanacearum, Xylella fastidiosa,* etc. (Compant et al. 2019).

9.5 Major Roles of the Soil Microbiome in the Environment

The impact of soil microbiome on the environment is what sustains the life forms on Earth. There is an in-depth connection between the microbiomes in the soil, with plants, animals, and humans (Banerjee and var der Heijden 2023). Several complex procedures in the environment are skilfully completed by these microscopic agents residing in the soil. Major processes in the environment like soil formation, biogeochemical processes—carbon cycle, nitrogen cycle, phosphorous cycle, and sulphur cycle (Naylor et al. 2022), and organic matter decomposition (Onyango et al. 2023) are microbiome mediated. Also, important processes like the decomposition of dead fungal mycelia are carried out by the bacterial species in the soil microbiome (Lladó et al. 2017). Bioremediation, the process of detoxifying and degrading the pollutants are mediated by the soil microbiome in a sustainable and eco-friendly manner (Prabhakaran et al. 2022).

Plant growth and development are largely influenced by the associated rhizosphere, which is mediating the processes of nitrogen fixation, nutrient mobilization, phytohormone production, stress response and control (Hakim et al.

2021). Soil microbiome also maintains the structural integrity of the soil, determining the soil fertility (Hartmann and Six 2022). The proper maintenance of soil health is ideal for agriculture. The health and prosperity of the agricultural ecosystems are maintained by the microbial community in the soil. Techniques are now being developed to manipulate the microbiome for agricultural advancements (Tosi et al. 2020). The use of microbiome-based agricultural management depends on the proper utilization of the plant microbiome interaction (Compant et al. 2019). The impact of the soil microbiome on the environment and the associated processes are numerous. The accomplishment of these tasks by the microbial community in the soil is what sustains and nourishes the other life forms on the planet.

9.6 Role of Anaerobes in Sustainable Agriculture

9.6.1 Anaerobic Bacteria in Sustainable Agriculture

The presence of anaerobes in the soil is dependent upon the moisture content. Excess soil moisture results in an anaerobic environment which is unfavourable for the growth of other aerobic microbes (Rui et al. 2023). Being adapted to temperature extremities, anaerobic bacteria are some of the common inhabitants of anoxic cold environments. Anaerobic bacteria isolated from such soils majorly belong to the genus *Clostridium* under the phylum Firmicutes. It is noted that they are also present in cold and flooded paddy fields, degrading carbohydrates, amino acids, and organic acids, and the degraded products are utilized as growth factors by other anaerobic bacteria in agricultural soil (Honma et al. 2023). An interesting observation made was that the agricultural practice of crop rotation reduced the frequency of anaerobic bacteria in the soil, whereas the practice of crop succession enhanced the survival of anaerobic bacteria as a result of soil being conditioned to lower oxygen levels (Cezar et al. 2021).

The rising population has impacted the agricultural sector tremendously. Apart from crop production, combating the plant-infesting pathogens is a major concern in agriculture. With the understanding of the hazards of using chemical based treatments, there has been a switch to opt for sustainable, environment-friendly agricultural practices. One of which is the Biological Soil Disinfestation (BSD) which uses the soil borne anaerobes (Patidar and Prakash 2022). This environmental biotechnology method uses organic materials to enhance the activity of the indigenous microbes to eliminate plant pathogens. Additionally, anaerobes used in this process releases acetate, butyrate (Zhou et al. 2021), and even enzymes, that terminates a wide range of pathogens (Ueki et al. 2018). This includes a list of nematodes, weeds, protozoans, and oomycetes, that infect crops (Priyashantha and Attanayake 2021). Studies have reported the anaerobic bacterial strain isolated from BSD soils belonged to the *Clostridium* genus, and were known to possess antifungal enzymes, capable of degrading the mycelial cells of fungal pathogens (Ueki et al. 2018).

Similar to BSD, is the sustainable agricultural technology of GET system (G-biomethane gas, E-renewable energy, T-*tanbo*-paddy field in Japanese), which generates and recovers methane from paddy fields. The analysis of the microbes involved in the GET system, revealed the dominance of Firmicutes, which performs functions like elimination of soil-borne pathogens and removal of nitrate, ultimately neutralizing the soil and inhibiting pathogens in the soil (Chen et al. 2022).

9.6.2 Methanogens Useful in Agriculture

Methanogens are anaerobic microbes that belong to the domain Archaea and generate methane gas as the final product of their metabolic processes (Li et al. 2021). The composition and distribution of methanogens in the soil varies from local to global levels, and is dependent on factors like pH, temperature, biogeochemical processes like carbon cycling, etc. (Li et al. 2021). Methanogens in the agricultural fields are the major producers of methane gas which contributes to global warming. But methane is also a necessity for plant growth and development, and response to abiotic stresses (Li et al. 2020).

Rice paddy fields are the major habitats for methanogens, and the coexistence of methanogenic taxa—*Methanocella*, *Methanosarcina Methanobacterium*, and *Methanothrix*, in the soil are linked to the high production of methane (Li et al. 2021). However, the use of fertilizer supplements in paddy fields, affects the composition of methanogens like *Methanosaetaceae*, *Methanosarcinaceae*, *Methanocellales Methanobacteriales*, and *Methanomicrobiales* (Barq et al. 2021). The biodegradable waste matter produced from agriculture are degraded by methanogens in biogas plants, which is now a widely used method for the generation of sustainable energy (Enzmann et al. 2018).

Currently, several innovative methods are employed to overcome plant diseases. In a recent study, the rice straighthead disease, resulting from the accumulation of dimethyl arsenic from microbe mediated methylation of arsenic in anoxic paddy fields, was shown to be alleviated using methanogens through demethylation (Chen et al. 2023).

9.7 Future Perspective

The triad of air, water, and soil maintains all life forms on Earth ranging from microbes to humans and animals. Human beings are the highest form of animals in the ecosystem, but our lives are incomplete without the minute and inconspicable life forms- the microorganisms. The essence and balance of the ecosystem relies on microbes. Soil is the largest known reservoir of microorganisms and the soil microbiome is an inevitable part in maintaining all the life forms on Earth. Therefore, exploring the soil microbiome is a potential way to utilize it for the betterment of mankind.

Over the years, the sky rocketed global population has increased the demands for food production tremendously. This is also accompanied by devastating impacts on the environment. Hence, it is important to balance the production with sustenance and this can be achieved through the novel sustainable agricultural practices. One way to achieve this, is through methods that exploit the microbes in an environment friendly manner. Since time immemorial, microorganisms have been in use in several ways, like fermentation, dairy production, pharmaceuticals, etc. Recent advancements in technologies have helped in exploiting microbes in agriculture, waste water treatment, production of biofuels, including biogas. The future beholds an even wider application of microbes in easing human lives.

Unlike the past few decades, we are now well equipped with high-end technologies, that help to analyze and understand the microbiome in depth. Understanding the genome and deciphering it will be the key to explore the applications of microbes to the fullest. The developments in the fields of molecular biology, genome sequencing, and metagenomics, will further aid in unravelling the microbiome, even from the previously inaccessible locations. The emerging field of omics will ease the study of microbial characters from genes to proteins. The information and interpretation thus gathered can empower the field of agriculture to a better tomorrow.

References

Abdul Rahman, N. S. N., Abdul Hamid, N. W. and Nadarajah, K. (2021). Effects of abiotic stress on soil microbiome. Int. J. Mol. Sci. 22: 9036. DOI:10.3390/ijms22169036.

Bahram, M., Hildebrand, F., Forslund, S. K., Anderson, J. L., Soudzilovskaia, N. A., Bodegom, P. M. et al. (2018). Structure and function of the global topsoil microbiome. Nature 560: 233–237. DOI: 10.1038/s41586-018-0386-6.

Banerjee, S. and var der Heijden, M. G. A. (2023). Soil microbiomes and one health. Nat. Rev. Microbiol. 21: 6–20. DOI: 10.1038/s41579-022-00779-w.

Barq, M. G., Hassan, M. M., Yasmin, H., Shahzad, A., Malik, N. H., Lorenz, N. et al. (2021). Variation in archaeal and bacterial community profiles and their functional metabolic predictions under the influence of pure and mixed fertilizers in paddy soil. Saudi J. Biol. Sci. 28: 6077–6085. DOI: 10.1016/j.sjbs.2021.08.091.

Bar-On, Y. M., Phillips, R. and Milo, R. (2018). The biomass distribution on Earth. Proc. Natl. Acad. Sci. U S A. 115: 6506–6511. DOI: 10.1073/pnas.1711842115.

Borowik, A. and Wyszkowska, J. (2016). Soil moisture as a factor affecting the microbiological and biochemical activity of soil. Plant Soil Environ. 62: 250–255. DOI: 10.17221/158/2016-PSE.

Cezar, R. M., Vezzani, F. M., Kaschuk, G., Balsanelli, E., de Souza, E. M., Vargas, L. K. et al. (2021). Crop rotation reduces the frequency of anaerobic soil bacteria in Red Latosol of Brazil. Braz. J. Microbiol. 52: 2169–2177. DOI: 10.1007/s42770-021-00578-0.

Chauhan, P., Sharma, N., Tapwal, A., Kumar, A., Verma, G. S., Meena, M. et al. (2023). Soil microbiome: diversity, benefits and interactions with plants. Sustainability 15: 14643. DOI: 10.3390/ su151914643.

Chen, C., Li, L., Wang, Y., Dong, X. and Zhao, F. J. (2023). Methylotrophic methanogens and bacteria synergistically demethylate dimethylarsenate in paddy soil and alleviate rice straighthead disease. ISME J. 17: 1851–1861. DOI: 10.1038/s41396-023-01498-7.

Chen, S., Hirano, T., Hayashi, Y. and Tamura, H. (2022). Biological soil disinfestation compatible with renewable energy production for sustainable agriculture. J. Pestic. Sci. 47: 111–117. DOI: 10.1584/ jpestics.D22-010.

Compant, S., Samad, A., Faist, H. and Sessitsch, A. (2019). A review on the plant microbiome: Ecology, functions, and emerging trends in microbial application. J. Adv. Res. 19: 29–37. DOI: 10.1016/j. jare.2019.03.004.

Do, T. T., Smyth, C., Crispie, F., Burgess, C., Brennan, F. and Walsh, F. (2023). Comparison of soil and grass microbiomes and resistomes reveals grass as a greater antimicrobial resistance reservoir than soil. Sci. Total Environ. 857: 159179. DOI: 10.1016/j.scitotenv.2022.159179.

Dong, L., Li, M. X., Li, S., Yue, L. X., Ali, M., Han, J. R. et al. (2024). Aridity drives the variability of desert soil microbiomes across north-western China. Sci. Total Environ. 907: 168048. DOI: 10.1016/j.scitotenv.2023.168048.

Enzmann, F., Mayer, F., Rother, M. and Holtmann, D. (2018). Methanogens: biochemical background and biotechnological applications. AMB Express 8. DOI: 10.1186/s13568-017-0531-x.

Faghihinia, M., Zou, Y., Chen, Z., Bai, Y., Li, W., Marrs, R. et al. (2020). Environmental drivers of grazing effects on arbuscular mycorrhizal fungi in grasslands. Appl. Soil Ecol. 153. DOI: 10.1016/j.apsoil.2020.103591.

Feng, H., Guo, J., Wang, W., Song, X. and Yu, S. (2019). Soil depth determines the composition and diversity of bacterial and archaeal communities in a poplar plantation. Forests 10: 550. DOI: 10.3390/f10070550.

Fiore-Donno, A. M., Richter-Heitmann, T., Degrune, F., Dumack, K., Regan, K. M., Marhan, S. et al. (2019). Functional traits and spatio-temporal structure of a major group of soil protists (Rhizaria: Cercozoa) in a temperate grassland. Front. Microbiol. 10: 1332. DOI: 10.3389/fmicb.2019.01332.

Hakim, S., Naqqash, T., Nawaz, M. S., Laraib, I., Siddique, M. J., Zia, R. et al. (2021). Rhizosphere engineering with plant growth-promoting microorganisms for agriculture and ecological sustainability. Front. Sustain. Food Syst. 5: 617157. DOI: 10.3389/fsufs.2021.617157.

Hao, J., Chai, Y. N., Lopes, L. D., Ordóñez, R. A., Wright, E. E., Archontoulis, S. et al. (2021). The effects of soil depth on the structure of microbial communities in agricultural soils in Iowa (United States). Appl. Environ. Microbiol. 87: e02673–20. DOI: 10.1128/AEM .02673-20.

Hartmann, M. and Six, J. (2023). Soil structure and microbiome functions in agroecosystems. Nat. Rev. Earth Environ. 4: 4–18. DOI:10.1038/s43017-022-00366-w.

Honma, S., Ueki, A., Ichimura, A., Suzuki, K., Kaku, N. and Ueki, K. (2023). Phylogeny and physiological diversity of cold-adapted anaerobic bacteria isolated from rice field soil in Japan. Microbes Environ. 38: ME22109. DOI: 10.1264/jsme2.ME22109.

Jansson, J. K. and Hofmockel, K. S. (2020). Soil microbiomes and climate change. Nat. Rev. Microbiol. 18: 35–46. DOI: 10.1038/s41579-019-0265-7.

Jiang, Y., Zhu, W., Zhu, K., Ge, Y., Li, W. and Liao, N. (2023). Similarities and differences in the microbial structure of surface soils of different vegetation types. PeerJ. 11: e16260. DOI: 10.7717/peerj.16260.

Lauber, C. L., Hamady, M., Knight, R. and Fierer, N. (2009). Pyrosequencing-based assessment of soil pH as a predictor of soil bacterial community structure at the continental scale. Appl. Environ. Microbiol. 75: 5111–5120. DOI: 10.1128/AEM.00335-09.

Li, D., Ni, H., Jiao, S., Lu, Y., Zhou, J., Sun, B. et al. (2021). Coexistence patterns of soil methanogens are closely tied to methane generation and community assembly in rice paddies. Microbiome. 9: 20. DOI: 10.1186/s40168-020-00978-8.

Li, L., Wei, S. and Shen, W. (2020). The role of methane in plant physiology: a review. Plant Cell Rep. 39: 171–179. DOI: 10.1007/s00299-019-02478-y.

Li, P., Li, W., Dumbrell, A. J., Liu, M., Li, G., Wu, M. et al. (2020). Spatial variation in soil fungal communities across paddy fields in Subtropical China. mSystems. 5: e00704–19. DOI: 10.1128/mSystems.00704-19.

Li, X., Qu, Z., Zhang, Y., Ge, Y. and Sun, H. (2022). Soil fungal community and potential function in different forest ecosystems. Diversity 14: 520. DOI: 10.3390/d14070520.

Lladó, S., López-Mondéjar, R. and Baldrian, P. (2017). Forest soil bacteria: diversity, involvement in ecosystem processes, and response to global change. Microbiol. Mol. Biol. Rev. 81: e00063–16. DOI:10.1128/MMBR.00063-16.

Lu, Z. and Imlay, J. A. (2021). When anaerobes encounter oxygen: mechanisms of oxygen toxicity, tolerance and defence. Nat. Rev. Microbiol. 19: 774–785. DOI: 10.1038/s41579-021-00583-y.

Mou, W., Kao, Y. T., Michard, E., Simon, A. A., Li, D., Wudick, M. M. et al. (2020). Ethylene-independent signaling by the ethylene precursor ACC in Arabidopsis ovular pollen tube attraction. Nat. Commun. 11: 4082. DOI: 10.1038/s41467-020-17819-9.

Naylor, D., McClure, R. and Jansson, J. (2022). Trends in microbial community composition and function by soil depth. Microorganisms 10: 540. DOI: 10.3390/ microorganisms10030540.

Onyango, L. A., Ngonga, F. A., Karanja, E. N., Kuja, J. O., Boga, H. I., Cowan, D. A. et al. (2023). The soil microbiomes of forest ecosystems in Kenya: their diversity and environmental drivers. Sci. Rep. 13: 7156. DOI: 10.1038/s41598-023-33993-4.

Patidar, P. and Prakash, T. (2022). Decoding the roles of extremophilic microbes in the anaerobic environments: Past, present, and future. Curr. Res. Microb. Sci. 3: 100146. DOI: 10.1016/j.crmicr.2022.100146.

Prabhakaran, A., Meenatchi, R., Pal, S., Hassan, S., Bramhachari, P. V., Kiran, G. S. et al. (2022). Soil microbiome: characteristics, impact of climate change and resilience. Understanding the Microbiome Interactions in Agriculture and the Environment 285–313. DOI: 10.1007/978-981-19-3696-8_15.

Priyashantha, A. K. H. and Attanayake, R. N. (2021). Can anaerobic soil disinfestation (ASD) be a game changer in tropical agriculture?. Pathogens 10: 133. DOI: 10.3390/pathogens10020133.

Rui, J., Zhao, Y., Cong, N., Wang, F., Li, C., Liu, X. et al. (2023). Elevational distribution and seasonal dynamics of alpine soil prokaryotic communities. Front. Microbiol. 14: 1280011. DOI: 10.3389/fmicb.2023.1280011.

Santalahti, M., Sun, H., Jumpponen, A., Pennanen, T. and Heinonsalo, J. (2016). Vertical and seasonal dynamics of fungal communities in boreal Scots pine forest soil. FEMS Microbiol. Ecol. 92. DOI: 10.1093/femsec/fiw170.

Taş, N., Prestat, E., Wang, S., Wu, Y., Ulrich, C., Kneafsey, T. et al. (2018). Landscape topography structures the soil microbiome in arctic polygonal tundra. Nat. Commun. 9: 777. DOI: 10.1038/s41467-018-03089-z.

Tosi, M., Mitter, E. K., Gaiero, J. and Dunfield, K. (2020). It takes three to tango: the importance of microbes, host plant, and soil management to elucidate manipulation strategies for the plant microbiome. Can. J. Microbiol. 66: 413–433. DOI: 10.1139/cjm-2020-0085.

Tripathi, B. M., Kim, M., Kim, Y., Byun, E., Yang, J. W., Ahn, J. et al. (2018). Variations in bacterial and archaeal communities along depth profiles of Alaskan soil cores. Sci. Rep. 8: 504. DOI: 10.1038/s41598-017-18777-x.

Ueki, A., Kaku, N. and Ueki, K. (2018). Role of anaerobic bacteria in biological soil disinfestation for elimination of soil-borne plant pathogens in agriculture. Appl. Microbiol. Biotechnol. 102: 6309–6318. DOI: 10.1007/s00253-018-9119-x.

van Leeuwen, J. P., Djukic, I., Bloem, J., Lehtinen, T., Hemerik, L., de Ruiter, P. C. et al. (2017). Effects of land use on soil microbial biomass, activity and community structure at different soil depths in the Danube floodplain. Eur. J. Soil Biol. 79: 14–20. DOI: 10.1016/j.ejsobi.2017.02.001.

Wang, M., Pu, W., Wang, S., Zeng, X., Sui, X. and Wang, X. (2023). pH-related changes in soil bacterial communities in the Sanjiang Plain, Northeast China. Microorganisms 11: 2950. DOI: 10:3390/microorganisms11122950.

Zehra, A., Raytekar, N. A., Meena, M. and Swapnil, P. (2021). Efficiency of microbial bio-agents as elicitors in plant defense mechanism under biotic stress: A review. Curr. Res. Microb. Sci. 2: 100054. DOI: 10.1016/j.crmicr.2021.100054.

Zhou, X., Lu, Y., Huang, L., Zhang, Q., Wang, X., Zhu, J. et al. (2021). Effect of pH on volatile fatty acid production and the microbial community during anaerobic digestion of Chinese cabbage waste. Bioresour. Technol. 336. DOI: 10.1016/j. biortech.2021.125338.

Index

For Product Safety Concerns and Information please contact our EU representative GPSR@taylorandfrancis.com Taylor & Francis Verlag GmbH, Kaufingerstraße 24, 80331 München, Germany